Help your Horse Live A Good, Long Life

Ten Simple Steps To Keep Your Horse Young

Karen E. N. Hayes, DVM, MS, PLLC

© Ironhorse Publishing
Hayden Lake, ID 2005

Help your Horse Live A Good, Long Life
Ten simple steps to keep your horse young

©2005 Ironhorse Publishing LLC
Photographs and illustrations by the author, with the exception of the swimming photograph on page 73, compliments of Callae Hyde.

Ironhorse Publishing LLC
21894 E. Hayden Lake Road
Hayden Lake, ID 83835 USA

Printed in USA

Library of Congress Cataloging-in-Publication Data

Help Your Horse Live A Good, Long Life
Ten Simple Steps To Keep Your Horse Young
by Karen E. N. Hayes, DVM, MS, PLLC — 1st edition.
ISBN 0-9747554-1-9
1. Horses. 2. Horses—Aging. 3. Horses—Health. 4. Horses—Care. I. Title.

HELP YOUR HORSE LIVE A GOOD, LONG LIFE
TEN SIMPLE STEPS TO KEEP YOUR HORSE YOUNG

TABLE OF CONTENTS

Help your Horse Live
A Good, Long Life

Ten Simple Steps To Keep Your Horse Young

ADAPTATION IS KEY.

INTRODUCTION

Aging is damage. Some of that damage is going to happen no matter what we do. But some of it is avoidable, and some is at least postponable. If we fail to help our horses avoid and postpone the damage that could have been avoided and postponed, then they're going to age prematurely. And that is *not* the same thing as living a shortened life. It's much worse.

That's because when your horse ages in a normal, healthy way, the damage that occurs in his body as he goes through the aging process occurs basically one cell at a time. It's a gradual, graceful, gentle process, with surprisingly little discomfort.

We often think of aging as a distinctly *uncomfortable* process, so how can we make it less so for our horses? Well, when the damage occurs a little bit at a time, spread out over time, your horse's bodily functions have a chance to *adapt*. Adaptation is the key to graceful aging—aging without pain, confusion, fear, disability, and social stress.

It's more complicated for people than it is for horses. For us, aging includes not only the physical changes but also emotional, spiritual, and money concerns, competitiveness, vanity, and other varied bits of baggage we've carried around since childhood.

With horses, there's a beautiful simplicity—adapting to the fact of getting older means adapting to physical and social changes, period. If you can help your horse do that, then even though, over time, he may have changed a lot in your eyes, *he* doesn't notice the changes, so as far as he's concerned, he's still functioning normally. When you're a prey animal, change is unsettling and stressful, while "business as usual" is blissfully secure. Day-to-day status quo means no reason to feel more vulnerable or anxious than usual. No undue stress.

Premature aging, on the other hand, is big damage, all at once. It can be very stressful. More importantly, your horse can't adapt to it—not 100 percent. So, suddenly, his day-to-day life is unsettlingly different, and the things he needs to do each day are difficult, possibly even painful—things such as competing successfully for food, getting to the water source, hearing the sounds of stealth, moving out of harm's way, finding a safe, sheltered place when the wind blows. As a result of his impairment, all the changes slated to occur throughout your horse's lifetime—which were supposed to happen down the road, one cell at a time—begin to topple early, in batches.

Each time this happens, it's another significant jolt, and it sets him back even more, and leaves him even more impaired. The whole process can become a tumbling snowball, and there isn't a thing about it that's gradual or gentle.

So our goal is not to prevent aging—that'd be nice, but it's not realistic. Our goal is to prevent *premature* aging. And the earlier in your horse's life we get started, the more successful we'll be. What we want is for our horses to go through their senior years in comfort, rather than discomfort. Enjoying their abilities, rather than dealing with disabilities. We want to elevate the quality of their lives, so they're living at the top of their game, from day one to the very last day. If, in the process, quantity wants to come along for the ride—a few extra years tacked on to their life span, *which commonly happens*—great. But that's not our primary goal. We want the quality first, and then, and only then, the extra years are more than welcome.

In the pages that follow, I'm going to give you what for me—for my own horses and for the horses I've cared for as a veterinarian—have been the most effective and practical ten steps to help horses avoid premature aging. And I'm going to start with the step that's the least sexy, the one you may be most tempted to skip. Hopefully, within the next couple of pages, I'll convince you that this step does indeed belong at the top of the top ten list.

STEP ONE
GET BLOOD TESTS

Every horse over the age of ten should get blood tests every year.

I'm not suggesting ten is old. It isn't. What I am suggesting is, many of the health issues that are common and easy to observe as horses get up in years are issues that got started much earlier, invisibly, when the horses were in their prime. If we can catch these issues early, we may be able to get them turned around, before they contribute to the aging process. Or, at least, we may be able to soften their effects, so your horse can adapt to them and live his normal life span—or more— with comfort and capability. Remember: *Quality* of life is as important as quantity.

These blood tests will help you identify not only what you need to do in order to meet your horse's needs, but also what you need to *not* do. Here's what I mean.

HELP YOUR HORSE LIVE A GOOD, LONG LIFE
STEP ONE

Senior horse care has become a popular topic. Every month, some horse magazine or cable TV show does a piece on how to take care of the senior horse. Local feed stores are hosting seminars on the subject, too. This is great. It's raising awareness and making a lot of information available that wasn't out there before. As a result, every horsekeeper might be able to do a better job of keeping horses happy and healthy well into their 20s and 30s.

But an inevitable and potentially dangerous side effect of all the press is that there's a lot of heavy marketing going on. These days, you're always hearing about all the stuff you "should" be giving your senior horse.

For example, "senior" feeds are selling like crazy. Many of them contain extra protein and fat. Some senior horses will benefit from this, but for more than a few, it's the *last* thing they need—for these horses, extra protein and fat will create problems, not solve them. The list of senior feeds and supplements available to you today, at your feed store and on the Internet, is growing fast. Which products, if any, should you consider? Which should you avoid?

That's where these blood tests come in. They can help you and your veterinarian sort out which feeds and supplements can help your horse stay younger, longer, and which ones you ought to avoid. This is not to say they're "bad" products, but the fact is, a large percentage of senior horses do have one or more health issues, and depending on what's going on inside your horse, some feeds and supplements will be good for him, and some won't.

STARTING AT 10 YEARS:
ANNUAL BLOOD TESTS

COMPLETE BLOOD COUNT (CBC)
"CHEM SCREEN"
SERUM INSULIN TEST

The first two tests, the CBC and chemistries screen, are offered by every veterinary laboratory on the continent as a package deal. It's the best deal in town. Your vet will come out and draw two small tubes of blood, submit them to the lab, you'll get the results tomorrow, and the whole thing will likely cost you under $100. The tests will look for a whole list of significant problems in your horse's internal organ systems, such as his kidneys, liver, heart muscle, protein levels, immune system, whether or not he's anemic, and so forth.

The serum insulin test will cost another $75 or so, depending on where you live. I recommend this test if your horse meets one or more of the following criteria:
1. He's an easy keeper.
2. He's ever had laminitis (or you're not sure, but you suspect it).

3. He has any of the symptoms of Equine Cushing's Disease (see page 119).
4. He looks like an old horse, even if he isn't.
5. Regardless of how he looks, he's 20 years of age or older.

The serum insulin test will tell you if your horse is insulin resistant. This is a condition that occurs in people as well. In fact, according to the Centers for Disease Control (CDC) and the World Health Organization (WHO), the developed world is in the midst of an epidemic of insulin resistance and its next-of-kin, diabetes. Insulin resistance in people is the gateway to diabetes. In horses, it's the gateway to laminitis and possibly equine Cushing's disease.

Horses don't get diabetes as a rule, but if you consider laminitis and Cushing's disease together, they form a syndrome that's very similar to diabetes. Both cause excessive urine production. Both are associated with circulatory problems in the feet. Both have an increase in body fat, and a characteristic way in which the body fat is laid down. Both are influenced heavily by diet. Both develop serious deficiencies in certain trace minerals. And, both cause premature aging.

In both humans and in horses, if you catch the problem in the insulin resistant stage, there's a very good chance you can turn it around and not only resolve the insulin resistance but also avoid going through the gateway to the big diagnosis that comes next. But, just as with diabetes in humans, once your horse has Cushing's disease, he's got it. We can prescribe treatment to help soften the effects of the symptoms, but we can't un-diagnose the disease. For both species, if you can catch the problem in the insulin resistant stage, you can dramatically reduce the risk of getting the bigger, "badder" diagnosis that follows.

The Arabian gelding in the photograph below is twelve years old. He feels good. He's active, strong, mischievous, and has a good appetite. He's obviously an easy keeper, so we opted for all three tests, and here's what we learned.

He has chronic hepatitis. And, he's insulin resistant.

The hepatitis is bad news, because we have no treatment for it. There is no medication we can give that will repair and replace the liver tissue that has degenerated. Technically the liver *can* regenerate itself, but unfortunately we don't know how to stimulate it to do that. When it happens, we call it "spontaneous healing" and we're grateful for it, but quite frankly it usually doesn't happen.

But there's good news: The owner of the gelding had no idea there was anything wrong with the horse, because he acts like he feels good. This means that despite having lost a significant amount of liver tissue, he still has some liver tissue left that's functioning well. So what we want to do is *take really good care* of that remaining liver tissue, so it'll last him a lifetime, at the top of his game.

We do this by making the liver's job easier. The liver is a multi-tasking organ, but one of its most burdensome jobs is to deal with the toxic by-products of protein and fat metabolism. So, what we want to do is make sure our little gelding is not getting any excess protein or fat in his diet, because any excesses in these areas will burden his liver and shorten its useful life. Remember, he's only twelve. The average life span of the average horse these days is 18 to 22 years—on the lower end of the scale for purebreds, and on the higher end for hybrids. For this to be a success story, we have to at *least* get this horse, who happens to be purebred, to 18 years of age, in comfort and capability.

If he were being fed a grain product, we could grab the label off the sack, take a look at the protein and fat levels, and make any necessary changes there, by choosing a different feed. But he isn't getting any grain at all—his diet consists entirely of local (orchard grass) pasture in the warm months and locally grown (orchard grass) hay in the cold months. So what we're going to do is set aside the liver issue for a moment and focus on his insulin resistance, because by doing so we'll do his liver a big favor.

If you've read any of the low-carb diet books, you know that if you're insulin resistant, you need to stay away from foods that have a high *glycemic index*. The glycemic index tells you how high, and how fast, a particular food causes blood sugar levels to rise. Research suggests that a high glycemic-index feed, which raises your horse's blood sugar levels really high, really fast, can eventually catch up with him and cause insulin resistance. And, most of the feeds we've been giving our horses are high glycemic. It stands to reason, then, that in a recent study, 70 percent of horses age 20 and over are insulin resistant. For reasons I'll explain in a moment, an insulin resistant horse who also has a liver problem will lose more liver tissue, faster, unless somebody fixes the glycemic index of his diet.

HELP YOUR HORSE LIVE A GOOD, LONG LIFE
STEP ONE

"**How come my horse is fat? All he gets is grass hay.**" We tend to think of grass hay as "safe." But there's a difference between native prairie grasses, and modern, cultivated grass. The horse evolved on native prairie grass, which is 75 percent carbs, but those carbs are *low glycemic*. Most cultivated grasses—grasses we plant and tend for the purpose of harvesting as livestock feed— have a high glycemic index.

Our patient's diet, orchard grass, is a high glycemic grass. So we took him off orchard grass pasture and hay, and put him on timothy grass hay, which is typically a medium-glycemic index hay. It's not the lowest glycemic index, but it's significantly lower than orchard grass, and perhaps as importantly, it's available where this horse lives, so we know the owner will probably follow our advice.

The payback from making this dietary change is twofold. It'll not only help him recover from insulin resistance, it'll also help his liver. Here's why.

When you're insulin resistant, your body's metabolism is confused. Instead of using carbs primarily for energy, and proteins primarily to support your immune system and build muscle, and fats to form the building blocks of hormones, your body takes everything you eat and turns it to fat. Not just any old fat, but an abnormal kind that behaves differently. It's deposited in a gloppy fashion, over the tailhead, in the groin, over the neck crest , and like a heavy cape over the shoulders.

In fact, it's possible for a horse to be *underweight* and still have this gloppy fat distribution—to have big, heavy deposits of fat in these four areas, and yet be ribby. That's because the *amount* of food isn't the problem. The *type* of food is. And when your horse becomes insulin resistant because he's been eating foods he was never designed to eat (such as high glycemic cultivated hay, and oats, and corn, and molasses, and so forth), this abnormal deposition of fat acts like a separate

organ, which perpetuates itself even when he isn't eating enough calories to sustain a normal body weight. And that fat is not only being glopped onto his body, it's also circulating in his blood, and going to his liver. When you have chronic hepatitis, you simply can't afford to let that happen.

Here's our boy six years later, at the age of 18, looking better than he did when he was half this age. All that gloppy fat is gone. Every day he gets after this is pure gravy. And, more importantly, he feels great.

Without the blood tests, we wouldn't have known about his liver, and his insulin resistance would likely have advanced to laminitis, which I consider to be the epitome of misery for horses. The insight we got from the blood tests gave us the chance to take action while he was still able to respond.

As a result, he has lived every day of the last six years in comfort, acting vibrant and energetic—his good quality of life has been not only maintained but actually boosted.

STEP TWO
FEED YOUR SENIOR HORSE AS THOUGH HE WERE A DIFFERENT SPECIES

When age catches up with a horse the way it has in the 26-year-old Half-Arabian mare pictured on the facing page, the standard response is to give an extra scoop of her usual feed, and—if she's really lucky—some corn oil. This is well intentioned, but it's unlikely to meet her needs. It may, in fact, speed up the rate at which she ages, particularly if she's developing a hidden health problem to go along with her advancing years. Here's why.

Your horse gets the "goodies" from the feed he eats by breaking it down. In order for his body to absorb nutrients from his feed, he has to break it down into *individual molecules*. Those are pretty small pieces. He does it in three steps.

1. First, he grinds it with his teeth.
2. Next, acids and enzymes in his stomach and upper intestinal tract chemically break those pieces into even smaller pieces.
3. And finally, good bacteria that live in his large bowel

break those little pieces into individual molecules of nutrients, which can then be absorbed by the lining of his intestinal tract.

The trouble is, as he ages, your horse digests feed less efficiently, because
- His teeth are wearing out and falling out,
- Those digestive acids and enzymes in his upper GI tract are becoming diluted, and
- The populations of good bacteria in his large bowel are becoming unstable.

As a result of these aging changes, the partially digested feed delivered to his intestinal bacteria isn't broken down as well as usual, which makes it more difficult for the bacteria to finish digesting it. They do the best they can with their unstable workforce, but the process will be inefficient and incomplete, and there'll be more "exhaust" (gas) produced. Any particles of food that aren't adequately broken down will go right on through, their nutrients unabsorbed. Hence, much of what your horse eats gets wasted, and his risk of gas colic goes up.

His risk of *impaction* colic is also raised, because the muscles that keep food moving through his gut are becoming sluggish. The longer his manure stays in the gut, the more moisture is pulled from it, and the drier it gets. The result: manure balls that are increasingly small, hard, dry, and prone to getting stuck.

To add insult to injury, there's also an increasing amount of scar tissue covering the lining of his intestinal tract, due to

- A lifetime of wear and tear,
- Worm damage (which I'll discuss further in Step Nine on page 80), and
- The cumulative effects over his lifetime of certain medications that are hard on the gut, such as "bute," flunixin ("Banamine" and generic forms), ibuprofen, naproxen, and so forth.

As a result, even well-digested feeds that are broken down into individual molecules may not get absorbed, because the windows of absorption are essentially boarded up with scar tissue.

So, the older your horse gets, the more he needs top quality ingredients that are easy to chew, easy to digest, easy to absorb, easy to pass, and produce little or no by products to burden his liver and kidneys. Feed that meets these requirements is said to have a high *biological availability*. Remember this term, because it's going to come up again.

Here's the bottom line as far as your older horse's diet is concerned. Younger horses thrive on tough, fibrous, prairie grass. For all the reasons discussed above, and a few others I haven't mentioned, feeding the older horse truly is like feeding a different species of animal. If all you want is a cheap way to fill him up, go ahead and give him an extra scoop of his usual feed. If you want to help keep him living younger, longer, you need to give him a feed that better meets his needs, and doesn't burden his system in the process. You need a top quality feed that was formulated with senior issues in mind, and only as much of that feed as necessary to meet his needs.

The good news is, today at every feed store you can buy "senior horse feed." This is a relative newcomer to the industry, and it offers an easy solution. But not so fast.

For about 70 percent of the senior horses in this country, many of the commercially prepared senior horse feeds are not the best choice—they may solve one problem while creating a few more. How can you tell whether a particular feed is right or wrong for your horse? The first thing you have to ask yourself is, "Is my horse insulin resistant?"

Thanks to Step One, you have that information. If your horse is insulin resistant, he needs a feed that has a *low glycemic index*. The trouble is, the USDA doesn't yet require feed manufacturers to add glycemic index information to the tag. But, this isn't rocket science. You can easily determine whether a particular feed is low glycemic. Here's how.

First, grab the label and look at the ingredients list. Look, in particular, at *the first five ingredients*. Now look at the list of low glycemic feeds in the box below my label.

If the *first two* ingredients on the feed sack label are on this low glycemic list, and *at least one* of the remaining three ingredients are on that list, then it's probably a pretty good, low glycemic index feed.

Now, I happen to believe that *all* horses should be on a diet that has a low glycemic index, even if they're not insulin resistant. You'll have to decide how you feel about that. My

SENIOR
"Complete Feed for Mature Horses"
Guaranteed Analysis

Crude protein, Min.	14%	Copper, min. ppm ... 30
Crude Fat, min.	4%	Zinc, min ppm ... 80
Crude Fiber, max.	18%	Selenium, min. ppm ... 0.30
Calcium, (Ca) min.	0.6%	Vit. A, min. IU/lb ... 3000
Calcium, (Ca) max.	0.9%	Vit. D, min. IU/lb ... 300
Phosphorous, Min.	0.4%	Vit. E, min. IU/lb ... 40

INGREDIENTS: Sun Cured Alfalfa Meal, Beet Pulp, Corn, Timothy Hay, Oats, Cane Molasses, Monocalcium Phosphate, Dicalcium Phosphate, Salt, Monosodium Phosphate, Vegetable Fat, L-Lysine, Manganous Sulfate, Zinc Sulfate, Vitamin A Acetate (Stability ... dl-Alpha Tocopherol (Source of Vitamin E), Vitamin D3 Supplement, Sodium ... Flavor, Choline Chloride, Copper Sulfate, Niacin, Folic Acid, Calcium ... d-Biotin, Thiamine Mono Nitrate, Riboflavin, Pyridoxine ... Cobalt Carbonate, Potassium Iodide, Live Yeast ... EA-SACC[1026]

Low Glycemic List

BEET PULP

SOYBEAN MEAL

LINSEED (FLAXSEED) MEAL

RICE BRAN

SOYBEAN HULLS

ALFALFA MEAL

WHEAT MIDDLINGS

WHEY PROTEIN

opinion is based on the premise that horses' digestive systems were designed to digest low glycemic feed. In their natural environment, wild horses live on native prairie grasses which are about 75 percent carbohydrates, but those carbs are *low glycemic*.

In humans with insulin resistance, there appears to be a rather poorly defined genetic influence—some people are more prone to developing insulin resistance than others—but diet has revealed itself to be a more powerful influence, because it can override the genes. And, according to type II diabetes researchers, as many as 60 percent of all people living in the developed world are "genetically susceptible." In other words, nobody seems to be immune. So, the more high-glycemic food you eat, the higher is your risk of becoming insulin resistant, and the more likely you are to develop diabetes, regardless of your genetic makeup.

Diet is why there's an epidemic of diabetes in our society, and that, I believe, is why there's an epidemic of insulin resistance in our horses, with as many as 70 percent of all horses age 20 and over afflicted. Seventy percent. When you get into numbers that high, you have to stop looking at insulin resistance as a disorder of the horse, and start looking at it as something wrong with the way we're taking care of them.

"But I don't want my horse to lose weight!" This is a common worry, when people hear me talk about putting all horses on low-glycemic feed. If a person is familiar with the glycemic index, it's usually because he or she has heard of so-called low-carb weight-loss diets such as Atkins® and South Beach®, which advocate low glycemic foods in the context of dropping extra pounds. It's logical to worry that if you put your horse on a low glycemic diet, he's going to lose weight, even if he doesn't need to, but that's not what happens. In fact, a horse can quite easily *gain* weight on a low glycemic diet, if weight gain is what he needs—and what he gains will be *healthy* weight.

Now, look at the guaranteed analysis on the label. Look, in particular, at the *crude protein* amount. Despite what you may have heard about older horses needing "lots of protein," the crude protein in your senior horse feed should be no higher than 14 percent. In fact, for many senior horses it should be lower than that—as low as 9 or 10 percent. The reason is, once again, if there's a problem in any of the major organs, too much protein becomes burdensome and can create health problems that'll hurt the quality and quantity of your horse's life. I'll go into more detail on this point in the next step.

Look also at the feed's *crude fat* content. It shouldn't be any higher than 4 percent. And, again, if your horse has any problems in his liver, we'd prefer it be even lower.

Finally, check the ratio of "minimum calcium" to "minimum phosphorus." According to experts, the calcium level should be no higher than one-and-a-half times the phosphorus. This is because senior horses have more difficulty absorbing phosphorus, and these two minerals play off each other—if one is too high, it drives the other one even lower. If your senior horse already has only a marginal amount of phosphorus in his system and you give him a feed that's too high in calcium, his phosphorus levels could get low enough to upset his electrolyte balance and create some health problems.

Once you've satisfied all these points—the glycemic index, the protein and fat levels, and the calcium-to-phosphorus ratio—you've got yourself a good *foundation* diet for your senior horse. By "foundation," I mean we don't necessarily expect this feed to meet all your horse's needs, but it will be a suitable baseline, to which we can add specific ingredients for his specific condition. Consult your veterinarian as well as the feeding instructions on the bag to determine how much of the feed your horse should get, according to his tooth power and body condition. As his teeth gradually fail, he'll need more of his senior complete feed to replace the forage he's increasingly unable to eat.

Unless your vet recommends otherwise, *always make forage available*, even if your horse only diddles with it. He was designed to graze 20 hours of every 24-hour day, so it's safe to say nibbling on forage is supposed to take up the bulk of his time. Depriving him of that opportunity could upset his contentment and encourage neurotic behavior.

HELP YOUR HORSE LIVE A GOOD, LONG LIFE
STEP TWO

I want to make one more point about the basic diet you use to feed your aging horse.

You may recall hearing on the news, years ago, about some research that showed a dramatic increase in health and longevity in laboratory animals that eat a "restricted calorie diet." There are reports of humans on a restricted calorie diet living an incredible ten to fifteen years—or more—longer than their peers who eat the "normal" way.

This is not new information. The first reports came out in the early part of the 20th century, based on some research done on monkeys. Since then, it's been repeated and confirmed in several other laboratory animals, and there is a growing population of humans purposely eating reduced calories for the purpose of enhancing their health and extending their lives by a significant number of years.

The rule goes like this. The average American woman eats around 1900 calories a day, and the average male eats around 2700. You're supposed to reduce your caloric intake by about 30 percent, but without reducing your intake of vitamins, minerals, and other essential nutrients. This means that the food you eat must be tip-top quality food, no wasted "empty" calories allowed. The payoff, by some accounts, can be a 30 percent increase in life span.

If you're interested, go to www.calorierestriction.org, where you'll find support, instructions, medical advice, and research references.

What does this have to do with your horse? The prevailing theory is that the reason we can live longer if we eat less food is that food, and its by-products, are toxic. That's why we were all born with a liver and a pair of kidneys—to deal with the toxic waste. Metabolizing food also generates toxic radicals, which accelerate the aging process.

Domestic American horses tend to be overfed and overweight. Consider—just consider—feeding your aging horse less *quantity*, but higher *quality* feed. In fact, consider doing this for yourself as well. That way, maybe you'll be around longer too, so you can ride your horse into the sunset.

STEP THREE
CUSTOMIZE WITH
THESE SUPPLEMENTS

In this step, you'll customize your senior horse's foundation diet to his individual needs by adding specific, top quality supplements. Following are the most common supplements prescribed for senior horses. I'll explain why your horse might need each one, how to give it, and where to get it.

Protein

Your horse may benefit from a protein supplement if

- His haircoat is dry and dull,
- His skin is dry and dandruff-y,
- He has poor quality hoof horn,
- Every time a horse in the neighborhood gets the sniffles your horse comes down with a full-blown respiratory infection (suggesting that his immune system is weakened), or
- His muscles seem to be getting smaller even though you exercise him regularly...

...and your veterinarian hasn't found any underlying health problems to explain any of these complaints.

You may wonder why I said the foundation diet should have a relatively low amount of protein in it, and now I'm suggesting some horses might need a protein supplement. Why not boost the protein content of the foundation diet, and skip the supplement? Three reasons.

First, most horses don't need extra protein. Second, some may actually do worse on it. And third, the protein found in most if not all foundation diets is the wrong kind for an older horse that needs extra protein. What do I mean by the "wrong kind" of protein?

Think of it this way. We add protein not because the horse needs *protein*, but because protein is made up of what the horse *really* needs—*amino acids*. Each protein source has a different array of amino acids in it.

Every species of mammal has its own list of what's called *essential* amino acids. These are amino acids the body can't manufacture on its own, and therefore must be provided in the diet. Lysine, for example, is one of the horse's essential amino acids. There are sure to be others, but they've not been identified because the research hasn't been done, and frankly probably won't ever get done. Therefore, we have to guess. The best way to make sure your senior horse is getting all the amino acids he needs is to feed a diet that has a broad spectrum of amino acids. The best way to do that is to make sure his protein comes from a variety of top-quality sources.

In most horse feeds, the major protein source is either alfalfa or soy. Of the two, alfalfa is more easily digested, so it's more desirable in my opinion. But the best protein, hands down, is whey protein.

Whey protein is almost never among the top five ingredients in commercial horse feeds because it's relatively expensive and would drive up the price of the feed. It contains the broadest spectrum of amino acids, including some that are not found in alfalfa or soy, so it's an excellent choice for supplementing your senior horse—it's more likely to fill any holes in his amino acid intake. Just as importantly, whey protein is *highly biologically available*—more so than alfalfa or soy—so it'll give your horse more nutritional benefits, more easily, without generating a lot of "garbage" to burden his liver and kidneys.

Whey protein is a fluffy white powder available from a number of sources, including feed mills and also Internet companies that sell specialty equine supplements. Whey protein from a human health food store will likely have vanilla or chocolate flavorings and sweeteners added, and it'll have a premium price tag, so I'd steer clear of those sources. The truth is, whey protein is whey protein, so the best way to shop around is to find the best price. I recommend that you check the American Dairy Associations's online library for sources by logging onto www.doitwithdairy.com and searching for companies that will sell bulk WPC34 (whey protein concentrate 34%)—that's the kind you want. The standard dose to start with is 1/4 cup mixed into the feed twice daily. A lesser product, which is still a good protein source but not quite as good, is "milk protein," which contains both whey protein and the protein known as casein. Milk protein is cheaper than whey protein, but for the purposes of supplementing a senior horse, whey protein is better quality and is definitely the way to go.

Fat

The most common reason to supplement your horse's fat intake is if he simply needs more calories (to gain weight) and you want to facilitate that without risking an increase in his intake of carbohydrates. Any vegetable oil will safely do the job; the maximum dose is a cup of oil in the morning and a cup at night. (Please don't feed your vegetarian horse animal fat.)

However, before you go out and buy a big jug of corn oil off the grocery store shelf, you should know that this is a golden opportunity to tip your horse's omega 3 / omega 6 fatty acid balance in favor of the naturally anti-inflammatory omega 3s. It's a rare senior horse who doesn't need this kind of help, as inflammatory diseases such as arthritis, laminitis, dermatitis, uveitis, myositis, etc. are almost routinely diagnosed in older horses and, for that matter, older people as well. This is a good reason to walk past the corn oil and pick up a higher quality oil that's got a richer omega-3 fatty acid balance, such as soybean oil.

Help Your Horse Live a Good, Long Life
Step Three

Not just any soybean oil will do. If the oil is exposed to heat in processing, its delicate omega 3 fatty acids will be destroyed. Therefore, you'll want *cold-pressed* oil, which is fragile and must be kept cool and protected from sunlight in order to keep it from spoiling (going rancid). The oils found on unrefrigerated grocery store shelves are heat processed, which gives them a longer shelf life, and leaves them nutritionally impoverished. They provide calories, nothing more.

A good vegetable oil that meets higher standards—extra calories, cold processed, and a rich balance of omega-3s— is a product called Cocosoya®, made of a blend of coconut oil and soy oil. High-end feed stores often carry it, but if yours doesn't, you can get it online at www.uckele.com. As an added bonus, if your horse is finicky, and/or if you're adding other supplements that are dusty or powdery or nasty tasting, Cocosoya has butterscotch flavoring added, which is so yummy that your horse probably won't mind even the yuckiest supplements. The product costs about the same as grocery store vegetable oil, but the drawback to ordering it yourself is that you'll have to pay shipping, which is about as much as the oil itself. So, if you can get your feed store to order it for you, that's a better way to go. But even if you have to order your own, it's well worth the expense.

If your horse doesn't need extra calories, you still might want to give him supplemental omega 3 fatty acids, to help protect him against (or help him deal with) any inflammatory condition. In my experience, every senior horse can use this kind of support, and your horse probably should get it. Omega 3 fatty acid supplementation will help his body keep inflammation from burning too hot or staying too long, and it may even make it possible for you to avoid the use of, or at least reduce the dosage of, anti-inflammatory medications, which can have serious side effects.

To give your horse a concentrated source of omega 3 fatty acids, without significantly increasing his fat or calorie intake, and/or without burdening an already troubled liver, the best choice is flaxseed oil. It won't take a significant amount of this oil to significantly boost your horse's omega 3 intake. You can buy it from specialty equine supplement companies, or at your local health food store. Once again, it must be kept cool and protected from sunlight, preferably *in the refrigerator.* Depending on your horse's size, the dose is 2.5 to 4 Tablespoons drizzled on his feed, twice a day. For more about inflammation and aging, see page 43.

Chondroprotectives
You might want to add chondroprotectives to your horse's feed, if he has arthritis. These are non-prescription medications that are supposed to promote cartilage healing and improve joint lubrication.

There's an ongoing debate about whether these products actually work. Research has shown repeatedly that the injectable, prescription-only chondroprotectives definitely do work—some are injected (by a veterinarian) into the muscle, some into a vein, and some directly into the joint. There hasn't been enough (and in many cases, *any*) good research to prove whether the oral, non-prescription products work, but most vets will tell you that they've seen cases where the oral products seemed to work, and other cases where they didn't. That's been my experience as well. At present, testimonials may be the best recommendation you'll get, and you'll have to use your judgment from there. In my personal and professional experience, the best approach seems to be to start with a course of the injectable kind, then maintain with a good quality oral product.

ADDING, SUBTRACTING, CHANGING
ANY CHANGES MADE IN A HORSE'S DIET MUST BE MADE SLOWLY AND GRADUALLY SO HIS INTESTINAL TRACT, AND SPECIFICALLY THE BENEFICIAL ORGANISMS LIVING IN IT, HAVE TIME TO ADAPT TO THE NEW FORMULA. REMEMBER THIS WHEN YOU ADD, SUBTRACT, SWITCH, OR IN ANY OTHER WAY CHANGE THE FORMULA IN YOUR HORSE'S DIET. TAKE A FULL MONTH TO EASE HIM INTO ANY NEW INGREDIENTS OR SUPPLEMENTS. OTHERWISE, HIS RISK OF INDIGESTION AND COLIC GOES SKY HIGH.

"Good quality" is the issue. There are about 150 different brands of oral chondroprotectives available for horses in this country, and as some of the key ingredients are expensive, the prices can cause sticker shock. Shop around, and read the labels to make sure a daily dose contains enough of the right ingredients. A lot of these products have just enough of the well-known, key ingredients to be able to include them on the label, but not enough to make a difference for your horse. In the sidebar below is a list of the components most often considered to be important, and the amounts needed, according to the latest clinical studies.

ORAL CHONDROPROTECTIVES
INGREDIENTS MOST OFTEN RECOMMENDED
GLUCOSAMINE HCL 5,000 MG TWICE DAILY
DEVIL'S CLAW EXTRACT 2,000 MG DAILY
VITAMIN C 4000 MG TWICE DAILY

OTHER INGREDIENTS OFTEN CITED:
CHONDROITIN SULFATE
BOSWELLIA SERRATA
CAT'S CLAW EXTRACT
YUCCA EXTRACT
COPPER
HORSETAIL
CETYL MYRISTOLEATE

<u>Digestive Aids</u>

If you feel your horse simply isn't getting enough out of the feed he eats, he may benefit from digestive aids, which include yeast fermentation products and probiotics (those beneficial bacteria mentioned earlier, which help digest food in the large intestine). There are several products available, made just for horses, that contain both the yeast and the probiotics; you'll probably be able to find one or two at your local feed store. Follow the directions on the label.

<u>Fiber</u>

Moisture in your horse's manure is affected by how quickly (or slowly) his food moves through his intestines—the slower it goes, the drier it gets. This is influenced by three things: how much fiber your horse eats, how much water he drinks, and how well his gut muscles work. We already know that in the aging horse, gut muscles tend to get sluggish. And, as your horse gets older and his teeth start wearing out, it's more difficult for him to handle his usual fiber source: hay. If he chews for a while and still hasn't managed to break the hay down sufficiently, he'll usually spit out the half-chewed wad rather than swallow it. As a result, his usual fiber intake goes onto the ground instead of into his gut.

Normal horse manure balls are plump and moist, and they tend to stick together somewhat when they hit the ground. The drier your horse's manure gets, the more it scatters when it lands. If his manure balls are small, dry, and hard, a fiber supplement is in order.

Psyllium is not a good fiber choice for long-term use because it can interfere with your horse's ability to absorb nutrients from his feed. In my opinion the best choice is whole flax seeds. A cup in the morning and a cup at night will add sufficient bulk to speed things up a bit in the intestinal tract. You can buy whole flax seeds at your local feed store; a 50-pound bag will cost about $25. These are naturally oily seeds and will go rancid if they're not stored in a cool dark place, protected from direct sunlight. Note that their tough hulls are likely to come through your horse's digestive tract unaltered. This means he won't get much nutritional value from the whole seeds; just the fiber effects. If you want him to get fiber *and* nutritional value, grind up the seeds just prior to feeding, using a coffee-bean grinder. If you're not going to feed this freshly ground meal right away, be sure to refrigerate it in the interim.

The other thing you must do to improve manure moisture is make sure your horse has free access to fresh water at all times, and also *plain white loose* salt. For some older horses, it's just too much work to get sufficient salt from a salt block—their tongues get tired and sore. If you've always used salt blocks, you may be surprised how fast a horse of any age will go through a supply of loose salt. You'll also notice an increase in the amount of water he drinks. Notice, too, that I said *plain white* salt. If the only salt you provide is mineralized, your horse will stop eating it when he can't tolerate any more mineral intake, which is likely to be long before he's satisfied his salt needs.

Vitamins

Most feeds sold by reputable feed companies contain enough of the vitamins we know to be important for younger horses, but a senior horse may benefit from additional vitamin E and vitamin C. These are powerful antioxidant vitamins, and much of the aging process is the result of oxidative damage.

Recent research has suggested that the horse can safely take as much as 10,000 IU vitamin E per day, which is significantly higher than the 1,000 IU government recommendation (which hasn't been updated for many years). However, this is controversial, and it's wise to be conservative. I give my senior horses 2,000 IU vitamin E per day, by squeezing the contents of a 1,000 IU capsule onto their feed twice daily.

Older nutrition texts indicate that horses manufacture their own vitamin C and don't need C supplementation. Recent studies have shown, however, that the levels of vitamin C in the bloodstream drop significantly as a horse gets older, suggesting that he may lose his ability to manufacture his own. It takes from 5 to 10 grams of C per day to bring the older horse's blood levels up to the younger horse's benchmark, so that is the dosage range I use and recommend. Start with 1 gram a day and gradually work up. If your horse develops loose stool, drop the dose back to a level he tolerates well.

Metabolic Minerals

If your horse is insulin resistant, has (or had) laminitis, or has been diagnosed with equine Cushing's disease, he may

41

benefit from supplemental "metabolic minerals." This comes on the heels of research on humans with insulin resistance and diabetes. Many of the associated health problems, such as the circulatory troubles that lead to amputations, wounds that won't heal, nerve problems, blindness from retinopathies, etc., seem to respond favorably to supplementation with magnesium, chromium, zinc, vanadium, sulfur, copper, and other minerals that apparently are "wasted" by the body during bouts of increased urination.

Limited clinical research suggests that horses with any of the metabolic disorders need this same kind of support, and when they get it, their condition improves. I have had good results with Advanced Biological Concept's 14-item "Free-Choice" mineral method. The A-B-C-Plus system also includes supplemental probiotics, to support the horse's ability to glean more minerals naturally, from his feed (www.a-b-c-plus.com). It's fascinating to watch a horse pick through the smorgasbord of minerals in the trays and choose what he wants. It's gratifying to see the results, within a matter of a few weeks. Another approach is to offer a balanced mixture of "horse minerals," such as "LMF Free-Choice Minerals" (www.lmffeeds.com). I've also used Uckele's "Glycocemic-EQ" (www.uckele.com), which contains many of the metabolic minerals. However, it's important to note that without the overall mineral balance, no targeted treatment will work optimally. Balance is critical. When the trace minerals are properly re-stocked in the body, all the bodily functions they govern simply work better.

STEP FOUR
FIGHT INFLAMMATION

Inflammation is probably the biggest culprit that will cause your horse to age prematurely. When I say "inflammation," I mean any of those conditions that end in –*itis,* including laminitis, myositis, arthritis, uveitis, gastritis, etc.

It really doesn't matter where the inflammation is in his body—if there's inflammation, there is the risk of premature aging. So when, for example, you look at the horse on the next page, who's having a flare-up of his arthritis in his right carpal joint, I want you to understand that his "knee" isn't the only focus of his problem. There are enzymes and polypeptides and substances such as c-reactive protein and interleukin and other chemicals coursing through his bloodstream that wouldn't be there if he didn't have the inflammation in that joint, and those chemicals have bodywide effects, aging him in general as well as stressing the affected joint.

The Friesian gelding in this photograph is 25 years of age. He had a strenuous performance career as a young adult—lots of pounding on those joints—so it's not a surprise that he has arthritis now. However, his arthritis is usually well managed with the use of omega-3 fatty acid supplementation, MSM (methylsulfanylmethenamine), oral glucosamine, vitamin C, and regular exercise. He is not generally a swaybacked horse, but he certainly is swaybacked in the photograph. It makes sense, if you think about the last time you were physically miserable from an injury or an illness. Were you concerned about standing up straight at the time? Probably not. You were preoccupied with your misery, and so you were slouching. This photograph shows one way a horse slouches. The muscles of his underbelly, which ordinarily support his back, have let go.

When you were a child, did anyone ever tell you, *If you keep making that face, it's going to stick!*? Likewise, if this horse's "slouch" is allowed to continue, before too long he will *be* a swaybacked horse permanently. That's because the weight of his belly (which is significant in a horse) is stretching the ligaments of his underbelly, as well as the ligaments holding his vertebrae in alignment, eventually to the point of no return.

He will not only *look* older, *he will age faster*, because in addition to the chemical effects of inflammation that are already circulating in his bloodstream, there will be inflammation and pain in his lumbosacral joint as a result of his new posture—the equine equivalent of our human "lower back" pain syndrome—and as a result he'll be less supple, less flexible, and more reluctant to move around. This, in turn, will aggravate his chronic arthritis. The swayback will also crowd his internal organs, which may interfere with digestion and set him up for colic. And, psychologically, people are less inclined to exercise a swaybacked horse, even though it'd be the best thing for him.

The bottom line is that the flare-up of arthritis in his knee (carpal joint) is affecting more than just his knee. There are four other horses in pasture with him, and they're all grazing. What's he doing? He's standing, not grazing, resting his sore leg, focusing on his discomfort. And, growing older by the minute. This is a good example of how difficult it is for a horse to adapt to the chunky (rather than gradual) damage associated with premature aging.

Inflammation is a funny thing. It causes heat, fever, swelling, redness, and pain, which sounds like something you'd want to avoid. But the truth is, there can be no healing without inflammation. Inflammation is a necessary evil which, in the final analysis, can do a body a lot of good. The trouble comes in when inflammation flares too hot, or burns too long and becomes chronic.

What's supposed to happen is, inflammation occurs when you need it (because of an injury or illness). It's supposed to stimulate healing, then shut off when healing is finished.

There are two groups of chemicals in your horse's body that regulate this process. One group consists of chemicals that are naturally *pro*-inflammatory—they stimulate inflammation to occur, in response to an illness or injury. They start the inflammatory cycle, and around and around it goes, causing heat, fever, swelling, redness, and pain. The inflammatory cycle is why, a day or two after getting a sliver in your fingertip, your fingertip is more hot, swollen, red, and painful than on the day you got the sliver. When healing is just about finished, the second group of chemicals—which are naturally *anti*-inflammatory—interrupt the inflammatory cycle. That's what's *supposed* to happen. The two chemical groups are supposed to be in balance with each other, so inflammation moves in when it's needed, and moves out when the job is done.

For a number of reasons, however, it's pretty common in an older horse for that inflammatory balance to get upset, and the pro-inflammatory chemicals to become dominant. That means the older horse is more likely to have an overreactive inflammatory response. This can result in allergies, autoimmune diseases, and/or inflammation that hangs around long after it's no longer needed, causing chronic inflammatory disorders such as arthritis.

When a horse has arthritis, what's the first thing we tend to do? Reach for the bute or some other sort of NSAID (non-steroidal anti-inflammatory drug). After all, we want our horses to be comfortable. Sounds like a reasonable thing to do, right? Well, there's a problem.

One of the major players in the pro-inflammatory group of chemicals is a chemical called cyclo-oxygenase 2, or cox-2. Most NSAID drugs work by blocking the action of cox-2, which makes the inflammatory cycle smaller and thereby helps relieve the heat, swelling, redness, and pain. So far so good. The problem is, most NSAIDS also contain a cox-1 inhibitor, and, wouldn't you know it, cox-1 is a major member of the naturally *anti*-inflammatory chemicals. So we're blocking inflammation in one location, and stimulating it elsewhere—in this case, in your horse's gut. It's a classic case of out-of-the-frying-pan, into-the-fire. And, remember, it doesn't matter where in the body the inflammation exists—wherever it is, it's contributing to bodywide, premature aging.

We're all so familiar and comfortable with bute and other NSAIDS, it's very common to hear someone say, "I love my horse so much, I'm just going to give him bute every day for the rest of his life, so his arthritis won't bother him." That's something that must be done with eyes wide open.

So what do we do? If your horse needs bute, give him bute. But at the same time, pull out all the stops and do everything possible to combat the inflammation in ways that don't cause more inflammation. Work with your veterinarian to give the lowest dose that'll do the job, for the shortest length of time, and then get him *off* the bute as soon as you can, while you support him in other, less harmful ways. Here are some examples of how you can do that.

First, supplement his diet with flaxseed oil (see page 36), because omega-3 fatty acids are the building blocks with which your horse will manufacture those natural anti-inflammatory chemicals. In many cases, omega 3s will allow you to reduce the dose of your horse's NSAID or steroid medication, and possibly even allow him to go off the medication entirely. Don't expect immediate, dramatic results. The effects are gradual and cumulative, but they're significant.

Once you're on an appropriate omega-3 supplement program, try as many other methods as you can.

• MSM is an effective anti-inflammatory in many cases, and I personally know a lot of veterinarians who use it on their own horses and recommend it for their patients.

• Acupuncture can be wonderful for relieving pain and inflammation; the only limiting factor is finding an experienced, credentialed veterinarian who routinely performs acupuncture on horses. Ask your regular veterinarian for suggestions, and visit the website of the American Academy of Veterinary Acupuncture at www.aava.org. Click on the "Directory" tab for a list of practitioners in your area.

• Don't forget the anti-inflammatory value of ice. For an excellent, flexible ice pack, mix 1 part rubbing alcohol with 5 parts tap water in a zipper-lock bag, stick it in a second bag to protect against leaks, and freeze. Massage it a bit to revive its flexibility before applying over a protective fabric layer on the affected body part. This ice pack is colder than standard ice, lasts longer, and chills the tissues better.

• For relief of the pain and inflammation (and damage) of arthritis, exercise is essential. (For more on exercise and arthritis, see page 59).

• Injectable, prescription-only chondroprotectives are almost certain to help if the inflammation is in a joint or tendon sheath.

• DMSO (dimethylsulfoxide) used topically is also a powerful anti-inflammatory.

• Physical therapy is also valuable.

• The new prescription product "Surpass" is an NSAID cream that reportedly stays where you put it, without causing bodywide side effects. Early reports from the veterinary field, including my own experience with it, indicate that it works well in selected cases, and seems to have little effect in others. It's certainly worth a trial.

• Topical capsaicin cream, which isn't anti-inflammatory, is a very effective pain reducer.

The point is, don't limit yourself to the "standard" anti-inflammatory tools. Be open minded, innovative, and expansive in your search for ways to keep your horse comfortable. This means that when you discuss treatment with your veterinarian, make it clear you're interested in ways of relieving your horse's inflammatory process without using harsh drugs that contribute to inflammation and premature aging. Without a headsup from you, your vet won't know where you stand on the issue. The bottom line: Take inflammation seriously, because it'll age your horse faster than most anything.

Back to our Friesian gelding. I did have to give him bute, preceding each dose with 12,000 mg of MSM to protect his

gut against the inflammatory effects of the bute. MSM is a mild, non-prescription, anti-inflammatory medication you can get at your local feed store. It has no known side effects. Two doses of bute, along with massage, ice, physical therapy and exercise, as well as his usual daily arthritis support, put out the major inflammatory fire, and from that point the support program—the chondroprotectives, the MSM, the flaxseed oil, and the regular exercise—picked up the slack and he's been fine. The photo above was taken three weeks after his flare-up.

We do what we can do. He's 26 now. I know that eventually some aspect of aging will catch up with him, and I accept that. But only when it's his time. Not a minute before.

STEP FIVE
NIX (OR FIX) THE STALL

If you believe that by bringing your horse into the warmth, comfort, and safety of a stall, you're doing him a favor, think again. There are many ways in which your horse's stall—if it's a typical one—is contributing to his aging process. Let's talk about the three biggest issues.

STALL ISSUE #1: IAD (INFLAMMATORY AIRWAY DISEASE)

IAD is inflammation in the tubes that carry air to and from your horse's lungs. Inflammation is heat, swelling, redness, and pain. When your horse's airways are inflamed, the tubes through which he breathes are reddened, swollen, irritated, and—because the lining of his airways secretes mucus when it's irritated—clogged with excess phlegm.

At the veterinary school at Michigan State University in the year 2003, researchers took elite performance horses who were doing well in their disciplines and had no cough, no runny nose, no fever, no outward sign of illness, and they "scoped" them (passed a flexible fiberoptic scope into their respiratory tracts) to see whether there was any evidence of IAD. What they expected to see was that the older horses in the group, who were around 15 years of age, would have some IAD as a result of the wear-and-tear of their careers, and the younger horses, who were around 5 years old, would have clean tracts. What they found instead was that *all* the horses had *significant* IAD, regardless of age. This was attributed not to their careers but to their *stalls*—from breathing the dust, allergens, and ammonia fumes there. As elite performance horses, you know these animals were well cared for, and they had nice stalls that were kept very clean, but you'd never know it from looking down their throats. Despite having no evidence of respiratory problems and bringing home ribbons, they had significant airway inflammation. Imagine how well they would perform if they could actually breathe.

This really isn't a big surprise. We've known for decades that the number one respiratory disease of older horses—heaves— is directly caused by the dust and allergens and ammonia in stalls. In fact, researchers often refer to heaves as *organic dust-induced asthma*. If you've ever had a heave-y horse, you know from experience that the only way you can help him breathe is to get him out of the stall.

At racetrack stables, respiratory infections run rampant. If a single horse comes down with the sniffles, everybody groans because they know that before long, the whole barn will be sick. About twenty years ago it became standard practice to vaccinate every horse at the racetrack for "rhino" and "flu" *every two months* to try and keep the outbreaks to a minimum. The results have been inconclusive. Sometimes it seems to help and sometimes it doesn't, but the practice of vaccinating every two months has continued, because the managers and trainers are afraid to stop—they can't afford to lose any ground and have their horses laid up any more than absolutely necessary.

The prevailing theory about why racetracks have so much respiratory disease is, 1) the bugs that cause the infection are highly contagious, and 2) the horses are under a lot of stress. Both are true. But there's a third factor: These facilities bed their stalls in straw or wood shavings, both of which have been implicated in IAD. If a horse already has inflammation in his respiratory tract, don't you think he's more susceptible to catching any respiratory bug that might be lurking in the neighborhood?

What is it that's so bad about straw and shavings? They contain organic dust and allergens that are highly irritating to the respiratory tract. We've been using them to bed our stalls for eons, for no other reason than it's the way it's always been done. We're caught in the belief system that stall bedding needs to be *soft and fluffy* so it'll be absorbent and comfie, and *disposable* so we can keep the stall clean. As long as we're stuck in that box, we won't be able to get out of this mess—respiratory ailments will continue to be a way of life at horse facilities

.

But there are sustainable (rather than disposable) ways to bed our horses that not only are absorbent, comfortable and clean, but which have many other benefits as well.

For example, for the last 15 years, out of frustration with the obvious drawbacks of "traditional" bedding, I've used what I call "STP" bedding in my stalls. That's Screened Topsoil and Peat. A mixture that's about 60 to 70% peat and 30 to 40% screened topsoil performs reasonably well for my stalls; the trick is to have enough peat to keep the soil from compacting too hard, and enough soil to make the mixture substantial enough to provide a supportive bed. Peat by itself is too fluffy; I'll explain why that's an issue in a moment. The makeup of topsoil varies a lot from yard to yard and region to region, so it's important to diddle with the mixture and perhaps even to try different soil amendments other than peat, such as well rotted and cured compost—anything that improves the tilth of soil is worth trying, However, stay away from sand, for obvious reasons (sand colic), as well as perlite or other volcanic glass materials (unless proven otherwise, they may contain free silica and could cause silicosis—black lung disease).

Maintenance of a sustainable bed such as STP is entirely different from maintenance of disposable bedding. You remove the manure daily. You do *not* dig out the urine spots.

Instead, you rototill the bed every day, to mix the wet spots with the dry spots and to aerate (infuse air into) the bed. The mixing process ensures that wet particles are in contact with dry particles, which speeds up the wicking and evaporation of the urine; it helps to soften and fluff the bed for comfort; and, by aerating the bed, in encourages the growth of "good" bacteria that will break down the urine *aerobically*. Aerobic breakdown of organic material such as urine is mostly smell-free. No ammonia. It's rather ironic. For sweet, fresh-smelling air, all you need to add to the bedding is... well, *air*.

When I first started this STP routine, I used my enormous, gas-powered garden rototiller, which was a chore even in my larger, 20 x 20 foot stalls. In typical 12 x 12 foot stalls, it'd be impossible to maneuver the thing effectively. Plus, it's a beast to start in the cold winter weather. Now I use an electric, lightweight tiller. There are several brands available—anywhere you can buy a lawn mower, you can find an electric tiller. They weigh about 20 pounds and cost about $300, and they work great. I've installed extension cords on retractable reels over each stall, so all I have to do each morning is reach up, pull down the cord, plug in the tiller, and away we go.

So, the daily routine is, let the horse out in the morning. Pick out the manure with a fine-tined manure fork. Bring in the tiller. Plug it in. Till the stall. I'm finished in less than 5 minutes. My manure pile is less than half its usual size because there's no urine-soaked bedding in it. I have fewer flies, because flies are attracted to the smell of ammonia. I'm not constantly buying disposable bedding to replace what would have been removed each day, because now I'm not removing any except whatever sticks to the manure. I merely top off the stall every month or so with an additional bale of peat or scoop of soil, depending on how well the bed is staying loose.

One word of caution: If your stall is small (the typical 12 feet x 12 feet), or if your horse is in the stall 24/7, sustainable bedding probably won't work for you. In order to handle the urine load, and to keep from compacting beyond the tiller's ability to re-fluff it each morning, it has to have some time off each day to recover—preferably 10 to 12 hours—and at least one dimension of the stall should be 20 feet (or, at the very least, the stall should have an attached, covered run to which your horse has constant access). If your setup doesn't have either of these deal-breakers and you're interested in trying STP bedding, I recommend you try it first in one stall, and experiment with the STP mix until you find the proportion of soil-to-peat that works best for the topsoil in your area. If, for whatever reason, you decide STP bedding is not for you, your used STP bedding can be used on your flower beds. (Don't use it in your kitchen garden unless it's composted first, in case there are fragments of raw manure in it.)

Most people love sustainable stall bedding, once they get used to how different it is. It's vastly better for the horse for more reasons than I can fit into this little book. It supports his feet better, keeps him (and his feet) drier, and he's a lot more likely to lie down and get rejuvenative sleep on this type of bed. And, the stall can be kept virtually dust-free—all you have to do is spray it with water during the hot/dry times of the year and till the moisture in. The bed should be just moist enough so that if you squeeze a handful, it'll form a ball that falls apart almost immediately and does not stick to your skin when you dump it out. Just as importantly, there's no ammonia smell, because aerobic maintenance encourages the growth of millions of beneficial aerobic bacteria—free barn help!

A few years ago, a commercial stall bedding product that's even better than STP came on the market. It's called Equidry Bedding®. Equidry is made of kilned clay that has been ground into amazingly hard little nuggets that resemble fish aquarium gravel or cat litter granules. However, unlike cat litter, Equidry doesn't crumble under the weight of the horse, and unlike STP it resists compacting, can be vacuumed and even power-washed, and can be disinfected. Horses love sleeping on it. It's not inherently dusty—it will become dusty from gathering dust in a dusty barn, but it can be cleaned. It molds to the shape of the horse's sole and thus provides excellent foot support, which should be beneficial for maintenance of a healthy "cup" (the concave shape of the healthy sole, which is necessary for good foot circulation, healthy hoof horn, and protection of the laminae). Maintenance of Equidry is exactly the same as STP.

Help Your Horse Live a Good, Long Life
STEP FIVE

As of this writing, Equidry is not available due to some administrative and logistical business issues, but I've been told by a reliable source that it will reappear on the market soon (check the updates page on www.theperfectstall.com). In the meantime, STP is a great way to get acquainted with the process of maintaining a sustainable stall bed.

Sustainable bedding of one form or another is going to be the bedding of the future, for a variety of reasons, including environmental issues, fire protection (most sustainable beds are not flammable; most disposable ones are), storage space, respiratory health, and costs of materials. For the purposes of this discussion, the main point is this: Properly maintained, sustainable bedding will not cause IAD. If you can eliminate IAD from your horse's respiratory future, you will have gone a long way toward protecting him against premature aging.

STALL ISSUE #2: STALL LIVING CAUSES ARTHRITIS

Look at your hand for a moment. Feel it. It's alive, warm, strong, hard working, and resilient. And, it heals when it's injured. It could not be or do any of those things without a healthy blood supply. Your bloodstream carries nutrients to the tissues of your hand, and carries away wastes. If you were to put a rubber band tightly around your wrist, within a minute your hand would start to turn blue, get cold, and go numb. Before very long, with its blood supply cut off, your hand would die. The beauty of your hand's circulation is, even if you're a complete couch potato, your hand gets its blood supply. It gets the nourishment it needs, and it gets its waste materials removed on a regular basis.

Now let's look at your horse's joints. The hardest working part of his joints is the cartilage. It's alive, strong, and resilient. It can heal when it's injured, although it's not particularly good at it. The reason: Joint cartilage has NO BLOOD SUPPLY. How does it stay alive, if it doesn't have a regular supply of nourishment, and regular removal of wastes?

It gets its nourishment and waste removal by marinating in *joint fluid*, also known as synovial fluid. This clear, pale yellow, viscous fluid is rich with nutrients. But, how does the joint fluid get to the deepest layers of cartilage?

The cartilage itself may look hard and shiny, but it actually has some "give"—some springiness—and it soaks up joint fluid, just as a sponge would do. When your horse bears weight, the cartilage compresses, and squeezes out the fluid it had absorbed. When he lifts weight off the joint, the cartilage expands and soaks up a fresh load of new synovial fluid.

So, when your horse walks around, with every single step he takes, his cartilage is compressing, then expanding, then compressing, then expanding... squeezing the fluid out, soaking it in, cleansing, nourishing, cleansing, nourishing.

Help Your Horse Live a Good, Long Life
STEP FIVE

His hoofbeats are the heart beat of his joints. If your horse is not moving, his joint cartilage is getting virtually no nourishment, and no waste removal.

Wild horses in their natural environment, on the prairie, have been observed to graze 20 hours of every 24-hour day. Grazing is not the same thing as sitting down at the table and eating a meal. It's a combination of eating and exercising. Nibble, step, nibble, step—nourishing and cleansing his joints while he eats. In a stall, with his feed brought to him, your horse is just standing there. If he stands there long enough, his joint cartilage goes hungry and starts to get sick from accumulated waste. Arthritis is the result.

The joints of modern day horses that have strenuous performance careers take a pounding. Nobody is surprised when, as these horses enter their teens, they develop arthritis. We find that to be logical. We blame it on wear and tear. It's true that a performance horse's joints suffer some damage during strenuous activity, but what's probably more damaging is going straight from that strenuous activity into a stall to stand for 20 hours or more, rather than into a pasture where he'd have the opportunity to step, and step, and step, and get his bruised cartilage nourished and cleaned.

Traditionally, we don't expect horses that don't have a performance career at all—who spend a significant amount of time living "the easy life" in a stall—to be just as likely to develop arthritis as the horses that work hard. But they are. Making your horse stand in a stall all day does the same thing to his joints as putting rubber bands around your wrists does to your hands. If your horse lives in a stall, it's pretty much guaranteed that *he will have arthritis* in his senior years. If he's already got arthritis, it's pretty much guaranteed that his arthritis will get worse in a stall. If he's a weekend warrior, brought out of his stall to over-exert once in a while, doing some degree of damage to his joint cartilage, then put back in the stall to stagnate, his arthritis will progress faster.

If there's any way you can get your horse out of his stall to live in pasture with free access to a run-in shelter, that's the ideal. If that's not an option, at least allow him to be outside, even if it's only in a paddock, during the day. And, *make his stall bigger*.

How much bigger? Obviously the bigger the better, but a study I did in the late 1970s suggested that there is something magical about 20 feet. In the study, horses were kept in stalls of various sizes, and the number of steps they took while in the stalls were counted. The stall sizes included 10 x 10 feet, 12 x 12 feet, 12 x 14 feet, 14 x 14 feet, 12 x 20 feet, and 20 x 20 feet. Statistically there was no difference in the number of steps taken by the horses in the stalls that had no 20-foot measure. The horses in the 12 x 20 and 20 x 20 stalls took roughly five times more steps than the horses in the smaller stalls—there seemed to be something magical about that 20-foot dimension.

Help Your Horse Live a Good, Long Life
STEP FIVE

If you can make your horse's stall at least 20 feet long or 20 feet wide, that's a great place to start. If you can do better than that, do. If you can attach an outdoor run, and leave the door always open, and put a cover over at least part of the run so he won't just stand in the doorway and look out at the rain, even better.

Then, spread out the points of interest. Put hay in one corner. Put grain in the opposite corner. And water in another corner. Set the stall up so he gets in the habit of using the whole area, and takes as many steps as possible. Consider these efforts to be your way of taking steps to keep your horse younger, longer.

STALL ISSUE #3: Stalls Disturb Sleep
What's supposed to happen is, your horse's major organs work hard while he's working, and rest and regenerate while he's asleep. That kind of battery recharging requires deep, REM (rapid eye movement) sleep. The only time your horse gets that kind of sleep is when he's sleeping flat out on his side for at least one *uninterrupted* hour. Race trainers know full well how important this is—a horse that gets good, rejuvenative sleep is a horse that runs faster and wins races. The trouble is, in the typical stall, very few horses get that kind of sleep.

One reason is, horses are prey animals that instinctively prefer to be in a herd, on the open prairie, where they can see all the way to the horizon without obstacles, hear the sounds of stealth, and rely on other herd members to help watch for boogeymen. In other words, as prey animals, before they're going to lie down, they have to feel safe and secure. If your horse doesn't feel safe and secure in his stall, only you can figure out why and fix it. It could be he needs to be able to see more of the landscape. It could be he needs other horses around him. It could be he needs someone to turn off that constantly playing radio so he can hear the sounds of the night and satisfy himself there are no lions, tigers, and bears nearby. Everybody's setup is different, and every horse is different, so you have to do that detective work yourself.

Once you've got the safety and security issue covered, consider the bedding. Stall bedding must beckon your horse to lie down. It has to have some characteristic he simply can't resist. I'm not talking about lying down, rolling, and getting right back up again. I'm talking about lying down, flat out on his side, and sleeping soundly.

The horses I've observed in large range areas seem to prefer to lie down and sleep in a grassy meadow—not the tall grass preferred by deer, but shorter grass that doesn't rustle in the wind or obscure the sleeping horses' vision. The ground is firm but yielding, so it cushions bones nicely. When they're getting up, which is an impressive athletic maneuver for any horse, they can plant their feet in the soft-but-firm turf and trust that they won't slip. When they're standing, the ground supports their feet—it gives a bit, so the walls of their hooves sink in a little, while the ground in the middle protrudes upward, supporting the soles. And, of course, the bed is dry, level, and free of rocks.

If a grassy meadow isn't an option, the next choice seems to be loose dirt, deep enough that it provides some cushioning. It allows them to plant their feet securely while rising, and it gives the added benefit of a skin treatment for protection against flies.

Neither of these choices is anything like the typical stall beddings. A 2004 study at a veterinary school in Denmark tested how long horses would lie down and remain on their sides in stalls bedded in straw versus shavings. They found that around half the horses never did lie down, and although those that did lie down did not get a full hour of rejuvenative sleep on either type of bedding, they slept three times longer on straw than on shavings.

This may be why race trainers prefer straw bedding at the track. Of course a stall bedded in straw is a labor-intensive chore to keep clean, but because of the sleep benefits, trainers may insist on it, so they can get more performance from their horses. The sustainable beddings I've already talked about—"STP" bedding (that's screened topsoil mixed with peat, see page 55) and Equidry bedding (see page 58)—are great for rejuvenative sleep. It seems that their firmness, quietness (no rustling), foot security (lack of slipperiness), and support are important features for the horse looking for an ideal sleeping place. Soft and fluffy bedding doesn't make much sense when you weigh in excess of 1,000 pounds and urinate a gallon at a time.

My own foundation mare, Murkje, is a ster Friesian mare I acquired as a 2-year-old, freshly imported from Holland. She has always been at the bottom of the herd hierarchy, so security is an issue for her, and I assumed this was why she would never lie down in her stall. In fact, the only time she did lie down, during the first 15 years I had her, was when she was in labor. The first night I put her on Equidry bedding, she slept "sitting up" on her sternum. The second night she laid flat out on her side. When I saw her on the security camera I thought she was sick (dead, actually) and ran out to check on her.

She was fine. More than fine. She was catching up on years of inadequate sleep. After an hour and a half of deep sleep, complete with a whinnying and-running dream, she got up, took a drink of water, urinated, and went back down to sleep for another hour and a half. And this has become her pattern, every night. Since that time, I've tried her in a stall bedded in STP, and she sleeps almost as well. She is now 19 years of age and looks like the clock has been turned back by about five years.

HELP YOUR HORSE LIVE A GOOD, LONG LIFE
STEP FIVE

For these and myriad other reasons, the bottom line is that life in the typical stall ages your horse. It ages him by creating inflammation in his respiratory tract. It ages him by creating inflammation in his joints. And it ages him by denying him the sleep he needs for normal regeneration.

If pasture living isn't an option, what can you do? You can make your horse's stall larger, so at least one dimension is at least 20 feet. You can improve the ventilation. You can get him out of his stall every day, preferably for ten hours or more. And, you can switch to a more supportive, inviting bedding that does not contribute to inflammatory airway disease and that encourages your horse to lie down and get rejuvenative sleep. Do these things and you'll have a good start on making your horse's living situation less likely to be a "longevity liability."

STEP SIX
INSIST ON DAILY EXERCISE

Research on humans has repeatedly shown that 30 minutes of moderate exercise, daily for six months, turns back the markers of aging by as much as ten years. Ten years!

There's no reason to expect regular exercise is any less important for horses. As we've already discussed, low-intensity movement helps protect your horse's joints against arthritis, which is a significant inflammatory factor in premature aging. Remember, in their natural setting, horses are up and moving around for 20 hours out of every 24-hour day.

In this chapter, we're not only talking about using regular movement to prevent premature aging. Rather, we want to ramp up the intensity, from mild to moderate, to get the heart pumping more than it would be doing during casual grazing. The aim is to boost vitality and longevity. The trick is to keep the exercise from becoming stressful. The point is not to condition or build your horse for an Olympic performance.

The point is to open the blood vessels, boost the metabolism, work the kinks out of the muscles, flex the hooves, support and challenge already existing athletic capabilities, change the scenery, and brighten the outlook. We're going for moderate movement and a good time, within the limits of your horse's physical abilities. What we're not going for is injuries or wear-and-tear damage, both of which can contribute to premature aging.

Wear-and-tear injuries are the result of repetitive movements (doing the same thing day in and day out). Acute injuries,

such as sprains and strains, occur when strength gets out of balance—the muscles and joints that always get used become dominant, and the muscles and joints that rarely get used become weak and easily overstressed. Ligament sprains and joint injuries are very common when muscle strength becomes unbalanced in this way. In humans, for example, the quadriceps muscle at the front of the thigh is typically overdeveloped,

71

while the hamstring muscle at the back of the thigh is typically underdeveloped. This is a prime setup for a knee injury.

Ligaments—those short, stout, strap-like bands that hold the bones in proper alignment within the joints—were never meant to be the sole means of support, especially not when the leg is under load. Ligaments are really only supposed to serve as an alignment *guideline*, while the real support is supposed to come from the muscles that surround the joint. The trouble is, the less active your horse is, the more his muscles diminish, and the more strain there is on his ligaments. If he's active, but he does the same kind of workout day after day, then some of his muscles diminish while others dominate, and his skeletal frame gets pulled out of alignment—an even greater risk of injury.

The best way to avoid these injuries in your horse is to keep him active, but mix it up. Work a different set of muscles every day. If you usually work your horse under saddle, then spend some time driving him. If you don't know how, find a good trainer and learn—it'll be good for you as well as for your horse. If you usually ride your horse in an arena, do some trail riding, and vice-versa. If you always work on the flat, do some hill work. If for some reason your horse can't carry or pull you, put on your hiking boots and take him for a walk. For a real thrill, find an experienced person to help you take your horse swimming—he already knows how; all you have to do is learn how to do it safely. It's a great workout with zero concussion on bones and joints. All of these variations will work your minds as well as your bodies, forge a tighter bond between you, and keep both of you interested, motivated, stimulated, and young.

STEP SEVEN
SCHEDULE DENTAL CHECKUPS

Your horse's teeth grow continuously and wear down from grinding. This means that as he enters his senior years, he's beginning to run out of tooth roots. As a result, his teeth begin to wear out... and fall out.

Regular dental care isn't going to save your horse's teeth, so it might be difficult to understand why dentistry is needed at all. The reason is, in the process of wearing out and falling out, his teeth can cause some inflammatory problems in the adjacent tissues, and as you already know, inflammation—no matter where it is—is a major contributor to premature aging.

One common problem in the older mouth is periodontal disease. In the photograph on the next page is a good example of this inflammatory gum process. Between the thick, yellow tartar and the gums of this horse are billions of bacteria which go straight into his bloodstream every time his lips or tongue run over his gums. As a general rule, heavy tartar buildup and periodontal disease are more common in horses that have any of the three big metabolic disorders: insulin resistance, laminitis, and/or Cushing's disease. (For more on these conditions, see pages 13, 24, 98, 110 and 119.) Infection in tooth roots can make drinking cold water and chewing extremely painful, either of which can lead to the biggest killer of older horses: colic. A good veterinary dentist will address all of these issues and help eliminate this significant aging influence that's often overlooked in older horses.

As the last inches of your horse's tooth roots grow out, the shape of his teeth change accordingly. His teeth become narrower,

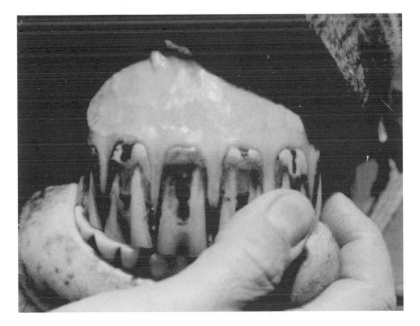

and as a result, spaces develop between them. Food begins to accumulate in those spaces, which can cause infection in the bony sockets where the roots are planted, as well as in the gum tissue. It's also not uncommon for sticks and twigs and other foreign objects to get stuck. That may seem like a small problem, but a tiny twig can do a lot of damage to the inside surface of your horse's cheeks, and not only severely crimp his desire to eat but also set him up for infection. These are jobs for the skilled veterinary dentist.

Because the teeth fall out one at time, spread out over a period of years, it's not uncommon for a tooth to suddenly be without its upper or lower occlusive partner—no opposing tooth to grind against. If this happens to your horse, the growth of the remaining tooth will accelerate, and there'll be no wear on its surface. As a result, it might actually get long enough to poke into the socket left behind by the tooth that fell out. This can be extremely painful, destructive, and an invitation for infection in his jawbone. It'll also, clearly, interfere with his ability to eat and chew. And, sometimes brittle old teeth break off before they fall out, which allows bacteria to go right into the jawbone.

Schedule veterinary dental checkups every six to twelve months for your horse regardless of his age and you'll help ensure that dental problems are never the cause of premature aging.

STEP EIGHT
VACCINATION DO's & DON'Ts

I'm often asked whether it's okay to give an older horse "a break" from annual vaccinations. The short answer is no.

One of the facets of aging is a gradual weakening of the immune system, so the protection your horse gets from vaccinations is important. But I do understand the desire to back off, and there is some merit to that notion.

Blood-donor horses that are used to produce hyperimmune serum get vaccinated as often as every month—much more often than the norm—so their blood will be rich in antibodies. The rather alarming truth is that many of those horses don't live past the age of ten. The constant barrage of immune complexes due to all the vaccinations can lead to kidney damage from accumulation of immune complex clusters clogging the kidney tubules, as well as a fatal liver condition known as liver amyloidosis. The point is not to alarm you and make you want to skip vaccinations altogether. I vaccinate my horses. The point is that, like most everything else in life, more is not necessarily a good thing—and frankly it's usually bad.

While vaccination will help keep your horse's immune system tuned up, it is possible to over-vaccinate, and there is good reason to be selective about what vaccinations you give your horse. This is not to say stop vaccinating. Rather, vaccinate him with the agents he needs, and back off on vaccines he doesn't need.

How do you decide what vaccinations are important for your horse? Talk with your veterinarian about what diseases are present in your neck of the woods, and what populations your horse rubs elbows with throughout the year. If he never leaves home, and he lives with horses that never leave home, then it's logical to think twice about vaccinating for diseases that have never been diagnosed in your part of the country. Depending on the epidemiology and effects of each disease, it might be okay to skip those shots, or to give them less often to your old guy.

Some vaccinations are non-negotiable. For example, definitely keep up with annual boosters of encephalitis and tetanus. These diseases are killers, and once-yearly boosters are very safe. And stay away from the use of tetanus antitoxin, which can cause a fatal liver disease called serum hepatitis. If your horse is well protected against tetanus thanks to annual vaccinations with tetanus toxoid, there should be no reason to use tetanus antitoxin, ever. The bottom line: Ask your vet to advise you, and make sure your horse gets the shots he needs, on schedule.

STEP NINE
DEWORMERS: CHOOSE WISELY

A lot of horsekeepers don't understand the difference between the two main types of deworming horses: using *purge* anthelmintics (where the horse is given a deworming agent, usually in paste form, every six to eight weeks) and *daily* anthelmintics (where a product containing the chemical pyrantel tartrate is given as a top dressing every day). It isn't just the schedule that differentiates these two options. The biggest difference is what each option was designed to accomplish.

In short, purge deworming was never meant to protect your horse against worms—it was meant to protect your horse's *premises* against becoming infested with worm *eggs*. In the long run, keeping the premises cleaner helps the general horse population 40 years from now, but it doesn't do much for your horse today. I'll explain in a moment.

When your horse picks up a worm egg, which hatches a larva (juvenile form of the worm), the larva doesn't go straight to the intestinal tract and stay there. Instead, it burrows out of your horse's digestive tract and begins to migrate through the organs and arteries in his body cavities. It tunnels through his liver, punches through his diaphragm, burrows through his lungs, punches its way into his windpipe, is coughed up and re-swallowed, and only then does it go to its preferred spot in his intestinal tract. Once there, it sets up housekeeping, matures to an adult, and prepares to lay eggs.

It's at *that* point that a purge dewormer is given—to kill the *adult* worm before it produces eggs that would contaminate the premises.

The relatively newer deworming agents that claim to kill larval forms of parasites are not the answer. Even if such a product were able to kill *all* the migrating larvae in your horse's body on the day the dewormer is given (which it does not claim), and even if it were able to convey residual effects to extend anti-larval protection for a while, it still can't protect your horse against all the larvae he picks up next week, and next month.

The larvae that migrate through your horse's body are the reason why it's common for a horse to come down with colic about two weeks after nibbling grass at the county fairgrounds, or at horse-friendly rest stops along the interstate highway system. Those little patches of ground are teeming with parasites from horses that passed through before, even though their manure is long gone and the grass is manicured. Within a couple weeks after grazing there, thanks to the larvae that are now mobilized and digging their tunnels, the horse's internal organs are besieged by inflammation and tissue damage. Even if he doesn't colic as a result (and many do), there are consequences down the road—scar tissue in his liver, in his lungs, in his blood vessels, and lining his intestinal tract. Over a lifetime, that damage accumulates to create some significant health problems, including the most common problem in older horses: a reduced ability to digest food and absorb nutrients. It's possible that sluggish gut muscles, which lead to impaction colic in older horses, are also the result of larval handiwork, clogging the arteries that feed the gut's muscles. Colic is the number one killer of the senior horse, and it's a particularly painful and violent way to go.

So now you know why the typical deworming program falls short of the mark—its stated purpose, eons ago, was to keep horse property, not horses themselves, from becoming infested with parasites. The fact that horse facilities *are* heavily infested today can't be blamed on the dewormers—they're highly effective agents, but their success depends entirely on timing. Deworm your horse a day or two late, and he will already have passed thousands of eggs into the environment before the dewormer kills the adult worms, and studies have repeatedly shown that many of those eggs can remain viable in the environment for as long as *40 years*. Even if every horse owner on the continent suddenly became highly vigilant about deworming their horses at precisely the right time, it'd be 40 years before those efforts would help a single horse.

Daily dewormer, on the other hand, is designed to kill those larvae *before* they embark on their journey through your horse's body. This approach spares your horse a lot of damage and also, by the way, protects the premises by preventing the larvae from maturing into egg-laying adults. In veterinary practice I've seen a major decrease in the number of emergency colic cases on farms that switched to daily dewormer.

Unfortunately the daily dewormer's chemical (pyrantel tartrate) does not kill all the important worms that afflict horses. (Nor, by the way, does any other dewormer of any type.) Pyrantel tartrate doesn't kill bots. And, I don't believe it kills tapeworms, either. There are research claims to the contrary, but those reports are based on fecal exams, and tapeworms don't shed their eggs in a regular, predictable fashion. If a horse isn't passing tapeworm segments, that doesn't mean he doesn't have tapeworms, in other words.

So, twice a year (in early spring and late fall), I give daily-dewormed horses a dose of a purge product that contains both ivermectin (which kills bots) and praziquantel (which kills tapeworms), and a day off from the daily dewormer.

If you're thinking that you dislike the idea of giving horses a deworming chemical every day or even every six weeks, I'm right with you on that. But, worms do an incredible amount of damage to animals that were never designed to live in confinement, where they're constantly exposed to equine feces and parasite eggs. And although there is an immune influence to parasite infestations—some horses are more "susceptible" to worms than others are—as your horse gets along in years, a worm-tough constitution will begin to work *against* him. That's because when a migrating larva gets attacked and killed by your horse's immune system en route to the intestinal tract, the dying and dead worm in, say, the middle of your horse's liver, creates a hotter, bigger inflammatory fire than do the live worms tunneling through his body. It seems that when it comes to intestinal parasites, domestic horses just can't win. They truly do need our thoughtful help.

Am I saying daily dewormer is better than purge dewormer? It depends. I definitely am saying the two are different and deserve some careful consideration before you decide which to use on your horse. Which is better for your horse depends on the worm load on your horse's premises, and on whether or not he has access to worm-contaminated areas. When you make your choice, make it an informed one. Know what each of the two options can and can't do for your horse, and don't just give your horse what everybody else gives theirs.

Help Your Horse Live a Good, Long Life
Step Nine

By the way, there is a prevailing legend floating around, that representatives at Pfizer (makers of the daily dewormer Strongid C) don't disclaim, indicating that there might be a negative interaction between pyrantel tartrate and the nonprescription anti-inflammatory medicine MSM (methylsul fanylmethenamine). If your horse takes MSM for arthritis, and you also give him daily dewormer, I recommend that you have your veterinarian draw a blood sample to check your horse for elevated liver enzymes. You know how Internet legends go—they spread like wildfire whether they're true or not. I can't confirm or refute the legend; it could be pure fiction, or it could indicate that in some (but not all) horses, there's an interaction. You're getting a chem screen done on your horse annually anyway (*a la* Step One, page 10), so be sure to partner with your vet to check those liver enzymes.

STEP TEN
PREVENT SOCIAL STRESS

Good health is more than just physical health. Social health is just as important for your horse as it is for you. If your horse does not have a firm position in his herd hierarchy, then there will be conflict. Conflict is stressful.

Ordinarily, every horse in a herd has a specific position, and he's very aware of which horses are above him in the hierarchy and which are below him. Being at the bottom of the totem pole is not necessarily a bad thing—that's a human judgment, not an equine one. As long as that bottom horse recognizes and accepts his position, everybody's happy. It's when the hierarchy is shifting that conflict, and stress, come into play.

The aging process often begins to affect a horse's ability to defend his social position. It could be the result of general

Help Your Horse Live a Good, Long Life
Step Ten

weakness, or a decline in limberness that makes him slower to move out of another horse's personal space. It could be that because of failing eyesight or hearing, some of the horses that are socially beneath him will start to be successful in invading his space and socially challenging him, in an effort to climb up the hierarchical ladder. If he yields to them, then the hierarchy shifts, and that's that. If he does not yield, physical altercations often occur. And, sadly, if the conflict isn't settled in good order, then often what happens is that the rest of the herd starts coming down on the combatants.

Researchers believe that's because in wild herds, a conflict distracts everybody from their number one job, which is to remain vigilant as a group, watching for predators. In other words, if two horses are constantly at each other, the whole herd is at risk because everybody is distracted, and their ordinarily well-coordinated defenses are upset. And so, the herd steps in to bring the conflict to a head. The end result can be that the older horse's life becomes very difficult. Suddenly everybody is picking on him, challenging his right to feed and water, refusing to allow him near. It seems cruel, but in the big picture, it resolves the conflict in a way that may be to the detriment of the aging individual, but good for the herd. The end comes sooner, rather than later, for him.

In the modern world, such herd dynamics don't affect herd survival, but they still have a major impact on health and survival of the individual horse that gets the short end of the social stick. In order to protect your horse against stress, which is a major contributor to premature aging, it's your job to monitor his social health and intervene before stress occurs. If he's getting into conflict with other herd members, step in. Either move the troublemakers to another pasture, or take your senior horse out and create a new herd that's kinder to him. Because even well-adjusted herds have conflict over feed, bring your senior horse into a protected area where he can eat without worry. Without you as his advocate, he's at the mercy of the other horses, and Nature's way is hardwired to be harsh.

SPECIAL SECTION: THE BIG THREE

WHAT TO DO IF YOU ACQUIRE A HORSE THAT HAS INSULIN RESISTANCE, ACUTE LAMINITIS, AND/OR EQUINE CUSHING'S DISEASE

If you have a horse that has fallen victim to one or more of the "Big Three" problems, you may have noticed that treatment recommendations made ten years ago for these conditions are now obsolete. Thanks to a surge in good-quality research in this area, funded in part by remarkable parallels to similar conditions in humans, we now have much more effective treatment for afflicted horses. This gives you real tools to use if your own horse falls victim, and creates a future for talented but suffering performance horses looking for a new owner — perhaps you, if you're willing and able to get treatment for a horse that's still sound and could be a great schooling mount.

Following is a summary of the newest information and recommendations for Insulin Resistance, Laminitis, and Equine Cushing's Disease. Remember, what we know about these disorders is a work in progress, so don't be surprised if a new edition of this little book shows up in a few years, with information that will supersede what's on these pages.

INSULIN RESISTANCE

Insulin resistance is a condition suffered by humans as well as horses (see page 13). At the present time, the best way to diagnose it is through the "serum insulin" blood test which measures serum insulin levels. Not all veterinary laboratories run this test, but your veterinarian can easily find a lab that will.

The sample must be drawn in the morning, before your horse eats any grain. If he has hay in his belly, that's acceptable, but if he gets any concentrate (grain) product as part of his daily ration, be sure to arrange for your vet to come and take the blood sample before feeding time. The sample must be spun immediately in a centrifuge and put on ice, so you'll need to provide an electrical outlet and possibly also some ice.

On page 12 in this book I mention five criteria for deciding whether your horse should get the serum insulin test. In most cases I've seen, the insulin resistant condition is so advanced that the horse "looks" insulin resistant, and all the blood test does is confirm the diagnosis. For example, if your horse is one of those that seems to "live on air," with a chubby body despite very little feed, that's a big tip-off. However, a lot of chubby horses would be sleek and smooth if they only got some exercise. The difference between the appearance of a horse that's just "couch potato" overweight and a horse that has insulin resistance is *the way the fat is distributed.*

The heavy, cresty neck, long recognized as a warning sign of impending laminitis, is the biggest outward indicator of insulin resistance. Over time, as the insulin resistant metabolism takes hold and begins to dominate, heavy deposits of fat also appear at the tailhead, in the groin area, and like a cape over the shoulders. Fat also begins depositing in the omentum— the lacy, cheesecloth-like apron that separates the internal abdominal organs from the lower belly wall. As omental fat accumulates, the belly of the horse becomes enlarged and rounded, like a big apple. This is similar to what happens to the belly of a human with insulin resistance.

This abnormal fat is virtually impossible to get rid of with a standard reducing diet. It behaves like a functional organ, perpetuating itself and driving the metabolism of the whole organism. In fact, I have seen insulin resistant horses that have the characteristic distribution of fat—the big round belly, the jiggly cape of fat on the shoulders, the heavy, crested neck— while they're actually ribby in general. Even when they're thin overall, their insulin-resistant fat distribution can prevail. It's that persistent.

If insulin resistance is so obvious in the horses pictured on the following pages, then what's the value of the blood test?

HEAVY DEPOSITS OF FAT OVER THE SHOULDERS OF THIS INSULIN RESISTANT HORSE HAVE A BULKY, CELLULITE-LIKE APPEARANCE.

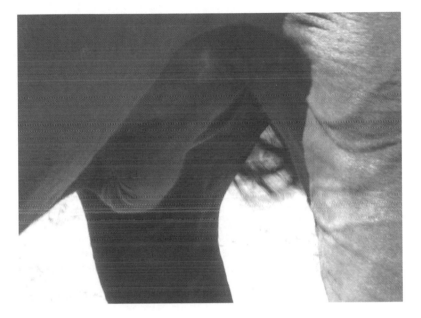

FAT IS ALSO DEPOSITED IN THE GROIN AREA, FILLING UP THE SHEATH OF THIS GELDING, GIVING IT AN INFLATED, PUCKERED APPEARANCE. IN A MARE, THE FAT IS DEPOSITED JUST IN FRONT OF THE UDDER.

It may allow you to get your horse diagnosed and get his insulin resistance turned around *before* he gains those extra 200 pounds or more, *before* his neck gets cresty, *before* his metabolic upset has him teetering on the edge of laminitis. The earlier insulin resistance is treated, the quicker and smoother will be the course of your horse's recovery, and the rosier life in general will be for him.

Researchers have looked for genes that make certain people more susceptible to insulin resistance, and their studies have given them more than they bargained for. They didn't find "a" gene that predisposes for insulin resistance. They found a whole gang of them. So many, in fact, that the odds of developing insulin resistance at some point in your life are higher than the odds of not developing insulin resistance. And, preliminary research in horses has concurred. The horses that don't develop the condition later in life are the exceptions. Seventy percent of all horses age 20 and over have it.

It's important to understand that the particular fat of an insulin resistant horse isn't just unattractive and embarrassing. It is a metabolic organ that drives your horse's whole-body metabolism. It is this fat that will, if left to its own devices, flip the chemical switches that will make your horse's feet break down with laminitis. This fat wreaks chemical, metabolic, and physical havoc over a horse's body in general. And it will not go away easily with a simple cutback in calories, because once

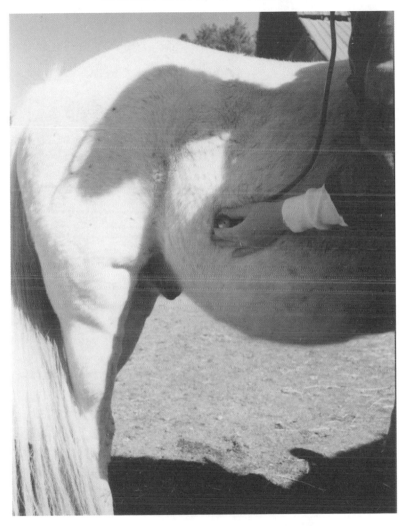

OMENTAL FAT GIVES THIS GELDING'S BELLY A ROUNDED, PROTRUDING, APPLE SHAPE.

it takes root and becomes a self-perpetuating organ, it does not respond very impressively to a "normal" weight-loss diet.

There are probably a number of ways in which insulin resistant fat can get established. What we know for sure is, chronic stress—stress that consistently plagues your horse day after day—causes increased secretion of the hormone *cortisol* by the adrenal glands. Some studies have suggested that the chronic stress that triggers and perpetuates insulin resistance for many horses may be nothing more than the metabolic instability caused by high-glycemic feed in the typical, modern-day, domestic horse's diet. Digesting his ration, in other words, may be a significant stress for your horse. If true, his ration must be terribly wrong for him.

Over time, the metabolism of the body that's exposed to too much cortisol becomes deranged and, among other things, causes the laying down of that special kind of fat which does not behave like "normal" fat. At some point, a threshold is reached, whereupon enough of this abnormal fat has accumulated to fuel a self-perpetuating vicious cycle. The metabolic changes ensure that there will be a constant supply of stress to the body, from the fat itself. It is at this point that ordinary approaches to fat reduction become ineffective, and the risk of laminitis looms large.

Because it isn't "just" fat, there are consequences to being insulin resistant. For a horse, those consequences include:

- Laminitis
- Increased oxidative stress (believed to accelerate the aging process), and
- possibly an increased risk of developing Equine Cushing's Disease (ECD)

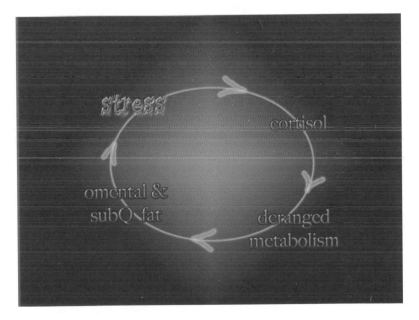

THE SELF PERPETUATING VICIOUS CYCLE THAT DRIVES THE INSULIN RESISTANT HORSE'S METABOLISM

The link between insulin resistance and Cushing's disease in horses isn't yet fully understood. Not every horse diagnosed with ECD is confirmed to have insulin resistance. But, there are some unavoidable parallels between the two. For example, horses that are "misdiagnosed" as having ECD, who actually have only insulin resistance, do improve when put on Pergolide (the current state-of-the-art, prescription medical treatment for ECD). If insulin-resistant (but not Cushing's) horses are switched to the state-of-the-art treatment for insulin resistance and taken *off* Pergolide, they continue to improve. However, horses that have ECD and are treated only for insulin resistance usually do not do well. So, we can't say that ECD is definitely the next step after insulin resistance, and we can't ignore the fact that the two disorders are related. We just haven't figured out quite how they're related, but that will come.

In horses as in humans, it is clear that diet, more so than genetics, plays a major role in the development of insulin resistance. And, as you'll see, diet also plays a major role in turning it around. Following is the game plan.

THE GAME PLAN TO BEAT INSULIN RESISTANCE

1. *Stay away from fat-enriched feed.* If you've looked into low-carb diets for yourself, you may know that a baked potato with added butter and sour cream has a lower glycemic index than a baked potato without the high-fat condiments. But hold on.

To add fat to a meal that's got a high glycemic index may be an acceptable way to lower the glycemic index *once in a while* if you're on a diet to drop fifteen pounds and are craving an ordinarily taboo potato. However, adding fat to your horse's *everyday* diet to lower the glycemic index of his ration might not be wise if he's seriously overweight, and particularly if he's insulin resistant. Added fat in your horse's ration, when his metabolism is already turning virtually everything he eats into fat, seems foolhardy, particularly when there are safer and less controversial ways to lower the glycemic index of the ration.

If your horse is getting a bagged feed of any sort, check the label for fat products in the ingredients list, and check the guaranteed analysis for crude fat content. If there are added fats or oils, and/or if the crude fat content is over 2 percent, I recommend you stop feeding this product.

2. *Switch your horse's ration to one that has a low glycemic index.* This isn't an effortless process, but it doesn't have to be difficult. Your ability to pull it off will depend on how flexible and committed you're willing to be to help your horse lose his abnormal fat and avoid, or recover from, insulin resistance.

We horsemen have gotten into the habit of insisting on "good" hay for our horses, without really knowing what "good" entails. Horses in the wild live, in their natural environment, on native prairie grass. This is a roughly 75 percent carbohydrate staple, but it's *low glycemic*. Most cultivated hay in this country is high glycemic, thanks to decades of well intentioned efforts in the agriculture industry to develop high-sugar forage for food-producing livestock.

The search for low-glycemic index hay for our horses is made more difficult by the fact that although certain types of grass have been shown to have a *trend* towards being higher or lower glycemic, there are some environmental and management factors that can cause a low glycemic grass to flip-flop and become a high glycemic one. So, use charts such as the one on the following page as guidelines, but don't take them as gospel.

Your task is to restrict your horse's access to any high-glycemic pasture, and replace his high-glycemic hay with hay that's got a low glycemic index. Go ahead and lock the pasture gate now, so you can have solid control over what your horse eats. But before you go in search of good hay, it's important for you to understand the environmental and management issues that will influence its quality for your horse.

Normally, the way grass grows, it collects "sugar" during the daylight hours, and gets taller at night. On a warm, sunny day, the sunlight interacts with chlorophyll (the pigment that makes the grass green) to fuel photosynthesis, which makes sugar. The grass does not grow during the day; it only manufactures

Glycemic Index Trends	
Higher Glycemic	Lower Glycemic
brome	bermuda
fescue	big bluegrass
oat hay	bluestem
orchard grass	crab grass
perennial rye	native prairie
quack grass	timothy
rye	

and accumulates sugar. The sugar-producing "factory" shuts down when the sun sets, and the accumulated sugar fuel is consumed at night to grow the grass taller.

If, during the night hours, when that sugar is supposed to be used up, to fuel growth of the grass, something interferes with that growth, then the grass won't grow and the sugar won't be used up. Even so, if the sun shines bright the next day, photosynthesis will resume, and more sugar will be added to the grass.

What could possibly interfere with the growth of grass that's well stocked with the fuel needed to support that growth? A number of things. Cold nighttime temperatures, for one. Poor soil, deficient in nutrients. Drought. A long heat wave. Too much rain. Overgrazing. Soil compaction. A pH imbalance of the soil. Trampling of the grass. Anything that stresses the system.

The longer one or more of these stresses persists, the shorter the grass will be, the higher its sugar content will be (because the sugar was never withdrawn from the grass's bank account to fund growth) and the higher will be its glycemic index.

On the other hand, the taller the grass, the higher its fiber content, and the more of its sugar will have been spent to build that fiber. And, the more of the remaining sugar will be enmeshed with the fiber in the grass. When sugar is enmeshed with fiber, it takes more work, and more time, for the body's digestive system to break down the fiber bonds and free the sugar from the mesh, and as a result the sugar is released more slowly into the system, resulting in a lower and slower blood sugar rise. A lower glycemic index, in other words.

It is a common misconception that "lush" grass is the culprit. There's no doubt a horse can get fat on lush grass. But, as you should now understand, crummy grass from a neglected field can be just as dangerous, if not more so, because it may have a higher glycemic index than the grass that was pampered, watered, fertilized, and stress-protected.

Just as the glycemic index varies with time of day, it also varies with the growth stage of the grass. The grass naturally stockpiles sugar in preparation for bloom and seed production. Grass eaten or cut prior to these energy-consuming processes will have a higher glycemic index than grass eaten or cut after it's matured and shed its seeds.

Thanks to the efforts of some dedicated researchers including Katherine Watts, who founded the website www.safergrass.org, it is no longer difficult to have hay tested for the qualities that affect its glycemic index. There's a dairy-oriented agricultural laboratory that she has groomed into a major force for equine-oriented feed testing: It's called Equi-Analytical (www.equi-analytical.com). The lab's staff will talk you through the hay-sampling process, tell you how to submit your sample, and help you and your veterinarian interpret the results.

WHAT ABOUT ALFALFA? It can be good for your horse, and it can be bad. Because alfalfa is usually a pampered crop lovingly tended by farmers who know what they're doing, while a lot of grass hay fields are dry (not irrigated) and harvested by less savvy hobbyists, there's a tendency for alfalfa to be less stressed. This could mean a lower glycemic index.

However, savvy farmers usually cut their alfalfa at a stage of growth that can maximize sugar content. Not good. And, some horses can't tolerate the high protein content that's often present in good quality alfalfa.

On the other hand, first-cutting alfalfa grown in the height of the growing season usually is taller, stemmier, more fibrous, and could be good, while fine, leafy, thready-stemmed, third-cutting, "dairy grade" alfalfa is probably the worst.

The bottom line on alfalfa is that it might be an improvement for your horse, if its glycemic index is significantly lower than that of the hay and pasture he's getting now, but it might create other problems because of alfalfa's other nutritional qualities, such as its typically high protein and calcium levels. You and your veterinarian will have to partner up on this issue to make the best decision for your horse. For me, a good rule of thumb is to go back to the notion of the wild horse in its natural environment, and the rough, fibrous prairie grass on which it thrived. I don't see a lot of alfalfa in that picture.

What if you're having a tough time finding low-glycemic hay for your horse, and your horse is in serious metabolic trouble? How do you feed him while you're shopping for hay?

Consider switching him to a grain-free, molasses-free complete feed made just for horses with insulin resistance. For horses living west of the Rockies, the feed company LMF (www.lmffeeds.com) makes just such a product, called LMF Low Carb Complete®. To use this for your horse means confining him to a drylot (so he can't graze high-glycemic grass) and feeding small amounts of the complete feed several times throughout the day. The danger is that people who actually have a life won't be able to get him fed little bits throughout the day and will, instead, give their horse his entire daily ration in two huge meals. For obvious reasons, this is not the best thing to do for an animal meant to graze for 20 hours every day. But there's an alternative.

The Pro-Feeder company (www.profeeder.com) makes an automated, durable, stainless steel feeder that will hold fifty pounds of product and mete it out in tiny, frequent, pre-programmed meals that will keep your horse busy eating small amounts spread out over a 24-hour period. It requires a power source—either an AC outlet or battery power—to run the program and the machine's mechanism. The company has been making this type of feeder for the dairy industry for decades and is just beginning to tap into the equine market. Their timing couldn't be better, because the number of insulin resistant horses in need of some major dietary help is huge. And, large meal size is one factor that can drive up the glycemic index of a ration. By using the automated feeder, you can keep the meals small and frequent, and the glycemic index of your horse's ration as low as possible.

3. *Soak while you shop.* It can take weeks to locate a viable hay source, get samples submitted, wait for the results, and get your new load of low-glycemic hay ordered and delivered. What should you feed your horse in the meantime, if you don't

want to go to a complete feed, but all you have in the barn is your usual, high-glycemic hay?

You can continue feeding your usual hay, but soak it first, in plain water. A one-hour soak in room-temperature water can lower the hay's sugar content by as much as half. The procedure is simple. Fill a clean tub with clean water. Dunk your horse's usual amount of hay for that particular meal. Let it soak for one hour, then drain, discard the water, and feed the hay immediately. (Don't soak later meals in advance to save time, and don't let soaked hay sit around before feeding it. It could become moldy.) This is a labor intensive process that you won't want to maintain for very long, nor will it be much fun in sub-zero weather. But it'll help you keep your horse fed while you're researching and securing your new feed.

4. *Balance the trace minerals* Add a free-choice, balanced source of minerals formulated strictly for horses, preferably formulated for horses with insulin resistance and associated metabolic upset. (For more on mineral balancing, see page 41)

Research on human diabetes has suggested that the miserable side-effects of being diabetic, such as the circulatory problems that can lead to non-healing wounds and foot amputations, the neuropathies, and the retinopathies, are not an inevitable part of being diabetic. Rather, they're the result of profound mineral wastage due to the high urine output associated with the diabetic condition. Studies have shown that patients with these diabetic "secondary disorders" can be seriously deficient in trace "metabolic" minerals such as magnesium, zinc, chromium, vanadium, and others.

Similarly, horses whose metabolic upset has caused an increase in urine output may have similar mineral imbalances and deficiencies that can cause changes in appetite, attitude, fat metabolism, thyroid function, and circulatory problems that make them more susceptible to laminitis. So, it makes sense to make sure your horse has access to these minerals in safe but generous amounts.

The Advanced Biological Concepts company (www.a-b-c-plus.com) has a clever mineral smorgasbord system that groups trace minerals and a digestive probiotic into an array of trays from which the horse can select what he needs. Although there are many researchers who doubt a horse can or will choose what he needs, there are a lot of clinical reports that suggest otherwise, including many from my own circle of clients and associates. Another option comes from the Uckele company (www.uckele.com) which sells a mixed trace-mineral product called "Glycocemic-EQ" which the company claims includes the minerals that are important to compensate for metabolic imbalances. In my experience, however, it's important to restore overall mineral balance first, before adding specific minerals with a specific goal. Once balance has been reestablished, everything seems to work better.

Although some have advocated fine-tuning a mineral supplement for each horse based on hair analysis (or at least hay analysis), I'm afraid that for most horse owners who don't necessarily have higher degrees in equine nutrition, interpreting the results and transcribing them into how much of this or that to give this could be an intimidating, time-consuming, frustrating approach requiring decisions they're not qualified to make. The truth is, many veterinarians aren't qualified to make those decisions either. However, if you want to give your horse a custom-designed mineral supplement, be sure to tap into reputable, ready-made resources that will help ensure your formula is safe. Another such resource is the LMF equine feed company's product, LMF Free-Choice Minerals (www.lmffeeds.com).

5. *Add antioxidants: Vitamin C and Vitamin E* The metabolic stress of insulin resistance has been shown to increase a horse's exposure to oxidative damage. Oxidative damage is what causes aging changes. The most powerful and well understood antioxidants for the horse are vitamins C and E. As mentioned in Chapter 3, it's a good idea to supplement these vitamins to prevent premature aging. You'll find dosage recommendations on page 41.

6. *Exercise regularly, if not acutely laminitic* Unless a horse has foot pain from acute laminitis, regular exercise at a mild to moderate intensity is an essential part of rekindling normal metabolism and improving overall health and resiliency. As mentioned in Chapter 6, the regularity and length of exercise

sessions is much more important than high intensity. In fact, high intensity exercise is a stress that should be avoided. Bear in mind that in their natural environment, wild horses graze (nibble, step, nibble, step) 20 hours out of every 24-hour day. Work up to an hour-long trail ride at a walk, with moderate hills. On alternate days, hire a trainer to help you and your horse learn to drive, and practice in the safety of an arena while you both build confidence. Consider swimming, if there's an equine pool or a pond or lake nearby. Take your horse camping—maybe even teach your horse to pack Ride another horse while you pony your horse. Always be looking for new, low-risk things your horse can learn to do, so you can avoid high-impact and repetitive activities that will add to his stress load.

7. *Stay as far away as possible from stress of any kind.* Remember, stress triggers the release of more cortisol in the body, which will fortify the vicious cycle that started your horse's metabolic upset. Look at stress as a multi-faceted issue that feeds on problems in the diet, environmental factors, social factors, and physical issues. Remember that for an animal that was designed to live on the open prairie with free access to shelter, stall confinement is as stressful as living out in the elements without shelter. Remember that for an animal that was designed to eat primarily native forage, a steady diet rich in starchy concentrates is a stress. Remember that a herd animal that isn't feeling up to par may have difficulty defending his position in the herd hierarchy and, as such, may

be a target for social stress from herd members. Be aware, vigilant, and proactive against stress from every angle.

8. *Support your horse's feet.* If he is insulin resistant, he is in a high-risk category for laminitis. Normally, the bottoms of his feet are concave, or cup-shaped. This configuration helps maintain good circulation in his feet and protects him against stone bruises and sole abscesses. If the soles of his feet are never supported, because he spends much of his time standing on soft, fluffy bedding or hard, unyielding ground, his risk of laminitis increases. The ideal substrate for him to stand on is a loose but substantial and somewhat conforming substance such as a grassy meadow; dry, loose soil; or a product such as Equidry (see page 58), all which will yield a little to the rims of his hoof walls and fill in the hollow underside of his foot to support weight there. This will preserve the cup and take some of the burden off his laminae.

LAMINITIS

The horse with acute laminitis is the epitome of misery. I have seen horses literally lie down and cry, for days on end, because of the inescapable pain associated with laminitis. In the bad old days, treating these cases was like groping in the dark, because a particular treatment worked in some instances and not in others, and the reasons were understood poorly, if at all. But we have come a long way, and although there is more to learn, laminitis is no longer the career-killer that it used to be.

Help Your Horse Live a Good, Long Life
Special Section: The Big Three

Researchers and clinicians are still arguing about some of the details in the treatment of laminitis, but they agree on many things, including rule number one: *Do Not Move Your Horse.* So, that's the first thing you must do, if you suspect your horse might be coming down with a case of laminitis. Confine him, in a stall bedded with a supportive bedding such as my "STP" mixture, or, if you can get it, Equidry Bedding. (See pages 55 and 58). If he's out in pasture, confine him where he stands, by having a helper halter and hold him, or by erecting corral panels around him. You'll be able to bring him into a stall, but not before you've wedged up his heels as described on page 117. You'll understand why in a moment.

Thirty years ago, the typical "sawhorse stance" of the laminitic horse, with front legs extended forward, was believed to be evidence that the horse was trying to take painful weight off his toes. This seemed to be confirmed by the fact that he reacted to hoof testers applied to the toe area. Veterinarians responded, with all good intentions, by lowering the horses' heels with a rasp, so they could bear more weight on their heels and less on those painful toes. It was the worst thing we could have done.

The illustration on the following page explains what we've learned, and why the state-of-the-art recommendations have changed.

One of the things happening inside the laminitic foot is that the delicate, corrugated membranes (the laminae) are being deprived of their blood supply. As a result, those membranes are sick and dying. Despite their delicate structure, they normally support the entire weight of the horse, by providing the connection between his coffin bone to the inside surface of his hoof. When the laminae sicken, their ability to hold that attachment begins to slip, and the coffin bone can begin to rip away from the hoof wall. This is a major source of the laminitic horse's pain. And, when that detachment begins, the weight of the horse, which is transmitted through the long bones of his leg, will drive that coffin bone right through the sole of the foot and into the ground, and all will be lost.

The DDF (deep digital flexor) tendon at the back of his leg normally attaches tautly to the back of the coffin bone, and its tug there will accelerate the ripping process. What we did in the past—rasping down the horses' heels in an attempt to relieve weight-bearing on their toes—tightened the DDF tendon and made things much worse. In retrospect, laminitis victims didn't stand with their front feet extended ahead of them in order to take weight off their toes; they stood that way in order to put some slack in their DDF tendons.

Another thing we've learned is that the shoeing approach we employed 30 years ago—the "heart bar" shoe—was the beginnings of a good idea that was taken out of the oven a bit too early. Because the weight of the horse is normally borne

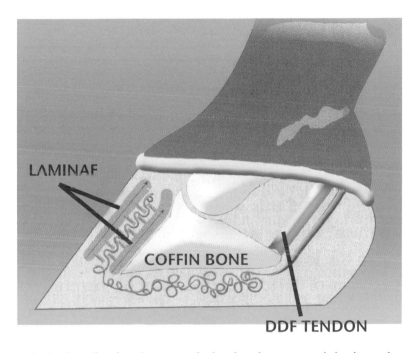

LAMINAE

COFFIN BONE

DDF TENDON

entirely by the laminae, and the laminae are sick, it makes sense to support the bottom of the foot, which is normally cup-shaped. The heart-bar shoe applied support only to the frog. The new concept, developed and well tested by researcher and farrier Gene Ovnicek, is to support the back two-thirds of the sole, with a substantial, supportive, yet somewhat yielding material such as construction styrofoam insulation. Ovnicek developed his Equine Digital Support System (EDSS) after studying the hoof structure of wild horses in their natural environment, and comparing it to the domestic horse in the healthy state as well as in various stages of laminitis. For a thorough review of the theory and practice of this approach, go to Ovnicek's website, www.hopeforsoundness.com. Send your farrier there, too, or ask Ovnicek for the names of farriers in your area who have trained in his approach—or, better yet, ask Ovnicek to fly to your place to treat your horse personally, because his approach is, in my opinion, the most likely to give you back your horse.

Another thing we've learned is that the circulatory events inside the foot as laminitis develops are similar to what happens in a human with a migraine headache. In the initial stages of a migraine, there is no pain, only a strange visual shimmering in the field of vision, called an aura, and usually an overall sense of feeling cold, with particularly cold hands. During that stage, the tiny blood vessels in the Saran-wrap-like covering over the brain have tightly constricted, limiting blood flow. Over a period of time, the aura gradually moves out of the visual field. It's at that point that the spasming blood vessels release their grip and dilate, and then the headache arrives. Pounding, debilitating, inescapable.

In the initial stages of laminitis, the tiny blood vessels that nourish the laminae spasm in the same way. This is called the *developmental* phase of laminitis. During this phase, there is no pain, no bounding digital pulse, and no abnormal heat in the hoof. In fact, the feet may actually be colder than usual. In the next, *acute*, phase of laminitis, the blood vessels expand. This is when the heat, pain, and bounding pulse appear. The delay between the developmental phase and the acute phase can vary, 24 to 72 hours. After the acute phase has been present for 48 hours, the case is considered *chronic*, and generally by this time, tearing of the coffin bone from the hoof wall has begun.

If the foot can be plunged in ice water during the developmental stage, research has shown that the acute phase can be completely avoided, and the laminitis can be halted before damage is done to the foot. The ice water inactivates destructive enzymes acting on the laminae, and preserves the delicate laminar tissues by chilling them to the point of lowered metabolism, so they can survive the spasming blood vessels and lack of blood supply. It also allows those spasming blood vessels to release the spasm and return to their normal state, without over-dilating.

However, timing may be critical. Some researchers believe that icing the laminitic foot once your horse is firmly in the *acute* phase, after the vessels have dilated and the foot has become hot and painful, may cause more tissue damage, rather than prevent it. Other researchers aren't so sure about that, but it's widely agreed that icing the foot in the *chronic* phase would be very damaging. The other problem is, as I've already mentioned, the *developmental* phase is basically symptom-free. So, how do you know if it's safe to apply ice?

In order to make this decision at the proper time, you must be aware that a laminitis trigger has occurred, or at least catch the case early in the acute phase. This is one reason why it's so vitally important that you be observant and spend time with your horse, and why you need to have a collaborative relationship with a good veterinarian and, just as importantly, a good farrier. If your horse has gotten into the grain bin and overeaten, that's all the clue you need. If he broke into the high-glycemic pasture and pigged out for an hour, there's another get-the-ice-boots clue.

If I have the slightest suspicion a horse might be showing early signs of laminitis, I strap the ice boots on right away. The good news is, recent research indicates that ice-bathing a healthy, pre-laminitis foot does not cause tissue damage and is quite well tolerated by the horse, even when the foot is kept submerged in ice water for 48 uninterrupted hours.

So, it's a good idea for every horse owner to have a pair of ice boots or a plan for how to ice their horse's feet if it were necessary to do so. If you suspect your horse might be at risk, consult your farrier and your veterinarian, and be prepared to act. When laminitis is suspected, every minute counts.

I mentioned earlier that you need to elevate your horse's heels if he is in any stage of laminitis. This will release tension on his DDF (deep digital flexor tendon), which will help prevent ripping of the laminae. Just a half an inch will do; more is not better. The photograph on the next page shows an easy way to achieve this.

A wooden wedge, as wide and long as his foot, duct-taped to the sole of each affected foot, will put some slack in the DDF while you wait for your veterinarian and farrier to arrive.

Because there are endotoxin, enzymatic and other chemical events that make the internal laminitis events destructive within the foot, aggressive treatment with medications known to block those chemical effects are also in order, even if your horse is not yet showing signs of discomfort. Your vet likely will prescribe the judicious use of pain killing NSAIDs (non-steroidal anti-inflammatory drugs) such as "bute" (phenylbutazone) and flunixin meglumine (Banamine®).

On top of these foot-focused treatments, it is wise to look at all but a few of the laminitis cases as metabolically caused. This means that if your horse has laminitis, it is more likely than not that he is insulin resistant. If true, then your treatment should include all the steps in the insulin resistance Game Plan (see page 98).

LAMINITIS TREATMENT PLAN

Developmental phase: ICE
Don't delay—Call your vet for help
Elevate the heels by 1/2 to 1 inch
Support the sole: EDSS
DO NOT MOVE THE HORSE
Follow Insulin Resistance plan

EQUINE CUSHING'S DISEASE

When I was in vet school in the 1970s, we were taught that if a horse lives long enough, he *will* get Cushing's disease—it's not negotiable, it goes with the territory of being an old horse. Today that notion is less certain, and there appears to be a link between equine Cushing's disease and insulin resistance. That link is not so direct that if a horse has insulin resistance he *will* develop Cushing's disease, nor is it true that *all* horses who have laboratory-confirmed Cushing's disease also have insulin resistance. Nevertheless, there does appear to be a definite pattern indicating that the two diseases are in cahoots.

Diagnosing Cushing's disease is usually based on a lab test called a low-dose dexamethasone suppression test (DST). It's actually a series of tests based on several blood samples taken in sequence, to measure the horse's response to an initial injection of the cortisol-like steroid, dexamethasone. The DST isn't perfect, but at the present time it's the best we have, and it's believed to be accurate about 85 percent of the time.

Some veterinarians, including myself, are a little queasy about prescribing this test in horses that are insulin resistant and therefore have higher-than-normal levels of the natural steroid hormone cortisol in their bloodstream. Horses with Cushing's disease also have elevated cortisol. Therefore, all of these horses are already in a high-risk category for laminitis, and there is some evidence that any additional steroid, including dexamethasone, could bring on this potentially devastating disease.

However, the only other reliable evidence of Cushing's disease is an external sign called *hirsutism*. Hirsutism means abnormal shagginess—the horse with Cushing's disease is shaggy, due to a longer-than-usual haircoat. As an indicator of Cushing's disease, hirsutism can be more reliable than the dexamethasone test in the sense that if your horse has hirsutism, he's got Cushing's. However, if your horse *doesn't* have hirsutism, that doesn't necessarily mean he doesn't have Cushing's, and sometimes when hirsutism does appear, it appears when the disease is already advanced. So, there's a place for the test, and the dose of dexamethasone is low, so it's worth the risk as long as you and your veterinarian have your eyes open and you're ready to pounce with treatment for laminitis if necessary.

Other signs of Cushing's disease are often present, but they're less consistent. Excessive water drinking and excessive urination may occur. And they may not. Excessive sweating occurs in some cases; lack of sweating occurs in others. Some horses with Cushing's disease act fatigued and depressed. Some act agitated. Some seem to behave normally, some seem vaguely "off." Some have decreased appetite. It's believed the reason for the variety of signs is the fact that the pituitary gland at the base of the horse's brain enlarges in Cushing's disease, and the longer the disease goes untreated, the larger the gland becomes. If it gets large enough to press on adjacent brain tissue, there will be outward signs. But those signs will vary according to what portion of the brain is being crowded.

The conventional treatment for Cushing's disease at the present state of the art is a drug called pergolide mesylate (Permax). It's a human drug, used in the treatment of Parkinson's disease. It works by promoting an increase in levels of the brain neurotransmitter *dopamine*. Decreases in dopamine levels are believed to be a trigger for the development of equine Cushing's disease. In response to the dopamine decline, the pituitary gland starts secreting increased levels of a hormone called ACTH (adrenocorticotrophic hormone) which goes to the adrenal glands and commands them to crank out ever-increasing levels of the stress hormone cortisol. The dopamine levels in the brain continue to drop, however, so the pituitary secretes more ACTH, more cortisol is produced, more ACTH, more cortisol, etc., and eventually the overworked pituitary gland begins to enlarge. We used to think this was a tumor on the pituitary gland, but now we know the gland is merely enlarging in an attempt to meet increasing demands for ACTH.

As you can see, equine Cushing's is a complicated disease with at least three factors contributing and causing signs— abnormally low levels of dopamine in the brain, abnormally elevated levels of cortisol in the circulation, and an enlarging pituitary gland which eventually gets large enough to cause problems by applying pressure on adjacent brain tissue. It is not yet fully understood what causes the drop in dopamine levels in the first place, and it is likely that this is where the link between Cushing's disease and insulin resistance may be hiding. Hopefully we'll have that information before long.

In the meantime, because of the possibility of that link existing, whether or not your horse has been diagnosed with insulin resistance, I recommend you treat him not only with pergolide but also with the insulin resistance game plan outlined on page 98. As I've mentioned before, I happen to believe all horses should be managed on a low glycemic diet anyway. As long as any dietary changes are made gradually, it's not going to hurt your horse to feed him the way he was originally designed to.

EARLY HIRSUTISM IN AN AGED SCHOOL PONY.

Pergolide is a prescription item, and there is a range of dosages used in the treatment of horses, based on body weight but also with the aim of starting with the lowest dose and gradually working up to the dose that seems to relieve the symptoms. Many Cushing's horses have a quick and gratifying response to pergolide treatment, with a decrease in water intake and urine output, a brightened outlook, and shedding of the abnormally long haircoat. Over time, that good response may wane. Whether this is because the treatment isn't fixing the underlying cause for the dopamine decline, or because advancing age is no longer interested in negotiating a truce, remains to be seen—we all eventually succumb to something. Pergolide is known to work better in early cases of equine Cushing's disease than in advanced cases, and this is probably true of most treatments for most ailments. The earlier, the better.

There is an herbal treatment made from chaste berry (*Vitex agnus-castus*) which, in European studies of alternative treatments, is believed to have a similar effect on dopamine levels as does pergolide. An equine formulation is available in the U. S. from a variety of suppliers on the Internet. The product is called *Evitex*. Results are mixed and there has not yet been a good study on the effectiveness of this treatment in horses. If your horse is in the early stages of Cushing's disease and you don't want to wait for those studies, this herbal treatment may be worth a try. Anecdotal accounts indicate that it, too, works better if used early in the course of the disease.

This means you've got a decision to make. If both the proven and the not-yet-proven herbal treatments both are said to work better in the earlier stages of the disease, it might be wise to give Pergolide a higher priority In other words, if you want to experiment with the herbal treatment and delay the use of Pergolide, don't wait too long.

In summary, if your horse is diagnosed with equine Cushing's disease, implement the insulin-resistant game plan (beginning on page 98) to minimize any dietary causes for elevated cortisol levels, and talk with your veterinarian about the pros and cons of starting your horse on pergolide and/or the herbal preparation Evitex.

About the Author

Karen E. N. Hayes is the award-winning author of the book *Hands-On Horse Care* and hundreds of magazine articles. *Help Your Horse Live A Good, Long Life* is her sixth horse-care book. Karen earned a Bachelor's degree in biology from Illinois State University in 1975, and her veterinary degree at the University of Illinois in 1979. She was in private practice for several years, joined the faculty at Wisconsin's veterinary school, earned a post-doctoral master's degree in equine reproduction there, then returned to private practice. She and her husband live in northern Idaho on a 100 acre farm (Ironhorse Friesians; www.ironhorsefriesians.com). Karen invites you to keep up with the latest horsekeeping innovations by checking www.theperfectstall.com.

OTHER BOOKS BY KAREN E. N. HAYES

• **The Complete Book of Foaling**.
A comprehensive, step-by-step, hands-on guide for foaling attendants. ISBN 0-87605-951-5

• **Emergency!** *The Active Horseman's Book of Emergency Care.*
Advanced first aid for the experienced performance horseman in a crisis, when immediate veterinary care is not available. Don't leave home without it. ISBN 0-939481-42-1

• **Hands-On Horse Care**.
Basic, detailed horse care, from daily observations to advanced procedures such as bandaging, medicating eyes, and giving shots, and a unique step-by step method for identifying the cause of the signposts a horse exhibits when injured, ill, lame, or just being a normal horse. In cooperation with the American Association of Equine Practitioners and winner of the 1997 American Horse Publications Book Award. ISBN 0-86573-861-0

• **Hands-On Senior Horse Care**.
How to take the best possible care of your aging horse. With the award-winning, signpost-oriented format, this book, co-written with *Horse & Rider* Magazine's former editor Sue Copeland, helps you decide whether your horse needs help, and walks you through the hows and whys. Packed with information, experience, and compassion. ISBN 192916411-4

• **The Perfect Stall**
Learn how the traditional stall falls short. Learn affordable and easy ways to improve the quality of the air your horse breathes, cut stall cleaning time and labor costs by more than half, save thousands in bedding costs every year, and improve comfort and safety for your horse. This book has started a horse-care revolution that is long overdue. ISBN 0-9747554-0-0